The publisher and the University of California Press Foundation gratefully acknowledge the generous support of the Chiang Ching-kuo Foundation for International Scholarly Exchange and the University Committee on Research in the Humanities and Social Sciences at Princeton University in making this book possible.

The publisher and the University of California Press Foundation also gratefully acknowledge the generous support of the Frank D. Graham Research Fund at Princeton University.

Continent in Dust

CRITICAL ENVIRONMENTS: NATURE, SCIENCE, AND POLITICS

Edited by Julie Guthman and Rebecca Lave

The Critical Environments series publishes books that explore the political forms of life and the ecologies that emerge from histories of capitalism, militarism, racism, colonialism, and more.

Continent in Dust

EXPERIMENTS IN A CHINESE
WEATHER SYSTEM

Jerry C. Zee

UNIVERSITY OF CALIFORNIA PRESS

University of California Press
Oakland, California

© 2021 by Jerry C. Zee

Library of Congress Cataloging-in-Publication Data

Title: Continent in dust : experiments in a Chinese weather system /
    Jerry C. Zee.
Other titles: Critical environments (Oakland, Calif.) ; 10.
Description: Oakland, California : University of California Press, [2021] |
    Series: Critical environments: nature, science, and politics ; 10 |
    Includes bibliographical references and index.
Identifiers: LCCN 2021024453 (print) | LCCN 2021024454 (ebook) |
    ISBN 9780520384088 (hardcover) | ISBN 9780520384095 (paperback) |
    ISBN 9780520384101 (ebook)
Subjects: LCSH: Dust storms—Political aspects—China—21st century.
Classification: LCC Q959.c6 z44 2021 (print) | LCC Q959.c6 (ebook) |
    DDC 363.739/20951—dc23
LC record available at https://lccn.loc.gov/2021024453
LC ebook record available at https://lccn.loc.gov/2021024454

28 27 26 25 24 23 22 21
10 9 8 7 6 5 4 3 2 1

For my grandparents

嚴華娟

徐一發

周欽德

孫桂中

So, Winds, you dare to mingle sky and land,
heave high such masses, without my command?

The Aeneid, Book 1, ll. 188–89

Wang looked up. The sky was obscured by a strangely
mottled layer of clouds. The clouds were made of dust,
stones, humans, and other odds and ends. The sun sparkled
behind them. In the far distance, Wang saw a long range of
transparent mountains also rising up. The mountains were
crystal clear, and changed shapes as they sparkled.

Liu Cixin, *The Three Body Problem* (2014, 222)

Contents

Illustrations

Acknowledgments

The bulk of the book was written and revised during extended stays in Vancouver, British Columbia, an eminently livable dystopia of settler colonial liberalism. It is also a place where many Asias tangle as speculative, flighty capital triples down to raise steel and glass towers of luxury investment properties in the extractive corridor and liquefaction zone that early Chinese migrant railroad builders called *Haam Sui Fow*, Saltwater City. Before and after all else, this place remains the unceded territories of the xʷməθkʷəy̓əm (Musqueam), Sḵwx̱wú7mesh (Squamish), and Səl̓íl̓witulh (Tsleil-Waututh) Nations. Following Iyko Day, "the interplay of Asian racialization, capitalism, and settler colonialism" (2016, 3) along the west coast of North America is this book's material and conceptual condition. I hope it is not its only horizon.

I am grateful to friends and mentors in China: Zheng Shaoxiong and Luo Hongguang, who supported my research through the Chinese Academy of Social Sciences; Altangarag; David Jiang and Kathi Zee Jiang, who always take care of me in Shanghai; and Chen Ting, Hilary Bauer, Tori Zwisler, ZeeZee Zhong, and Dan Guttman at Shanghai Roots & Shoots, who first brought me tree-planting in Inner Mongolia. Lee Seonhwa was my first friend in Inner Mongolia and facilitated my research in Seoul.

I cannot begin to express my gratitude to the many people in China who go by pseudonyms in this book, especially the people I call Tingting, Li Ming, and Xia Jie.

The work that became this book was financially supported by UC Berkeley's Institute for East Asian Studies, the UC Berkeley Center for Chinese Studies, FLAS, the Wenner-Gren Foundation, and the now sadly defunct (defunded) UC Pacific Rim Research Program. The University of California Humanities Research Institute and the Hellman Foundation supported a manuscript workshop in February of 2020. The American Council of Learned Societies and Mellon Foundation provided time to complete the dissertation that is an embryonic version of this book, and a Hunt Fellowship from the Wenner-Gren Foundation provided me valuable time to finish and revise the manuscript. The Chiang Ching-Kuo Foundation and the Princeton University Committee on Research in the Humanities and Social Sciences provided generous subventions to fund publication.

Before I knew better, Miyako Inoue, Paulla Ebron, and Sylvia Yanagisako encouraged me to pursue anthropology. They nurtured my curiosity and overlooked my blundering naivete as an undergraduate. Cori Hayden, Alexei Yurchak, You-Tien Hsing, and Jake Kosek served on my various graduate committees at Berkeley and waited through my rambling with patience. Aihwa Ong, my advisor at UC Berkeley, has supported me since before I began my doctoral work and, now, long after. She has always been ready with doting care, razor wit, and the exhortation to hurry up and say why it matters. She has reassembled my world more than once.

Ned Garrett's tireless and good-humored work in the UC Berkeley Anthropology Department made this and so many other works possible. My gratitude goes to many friends and teachers at Berkeley: Jia-Ching Chen, Nate Coben (who furnished the book with its *Aeneid* epigraph), Gabriel Coren, Shannon Cram, Joshua Craze, Rachel Cypher, Lindsey Dillon, Sam Dubal, Ugo Edu, Ruth Goldstein, Drew Halley, Monica Huerta, Janelle Lamoreaux, Heather Mellquist Lehto, Amelia Moore, Emily Ng, Milad Odabei, Jason Price, Raphaëlle Rabanes, Na'amah Razon, Takeo Rivera, Daromir Rudnyckyj, Marlee Jo Tichenor, Bharat Jayram Venkat, Max Woodworth, Hentyle Yapp, Gabe Yoon-Milner, Irene Yoon-Milner, and so many others. Pheng Cheah, Lawrence Cohen, Mariane Ferme,

Anne-Lise François, Liu Xin, Don Moore, and Stefania Pandolfo were generous teachers.

During a postdoc at UC Davis, the STS community provided valuable feedback and gave me the courage and confidence to reconceive and rewrite the project from the ground up. Special thanks to Joe Dumit and Marisol de la Cadena. Tim Choy served as my postdoc mentor. He became a trusted friend, interlocutor, and coconspirator during and after my postdoc at UC Davis. He made me slower, funnier, and kinder. I mean it when I say: this book is impossible without our friendship. I encourage you to read it alongside his forthcoming work on breathing and atmospheres as one half of an unplanned duograph (see Howe 2019, Boyer 2019).

At UC Santa Cruz, Andrew Mathews, Anna Tsing, and the students of the Landscape Lab demanded that I cultivate new arts of noticing. I cannot understate their impact on my thinking and my understanding of being a scholar and colleague. micha cárdenas, AM Darke, Muriam Haleh Davis, Amy Mihyang Ginther, Camilla Hawthorne, Mark Massoud, Adam Millard-Ball, Thomas Serres, and Katy Seto buoyed me. Thank you to Neel Ahuja, Mark Anderson, Lissa Caldwell, Nancy Chen, Sylvanna Falcón, Gail Hershatter, Emily Honig, Christine Hong, Megan Moodie, Laurie Palmer, Ben Read, and Peter Weiss-Penzias for intellectual inspiration and unwavering support.

I am incredibly grateful for the Princeton Department of Anthropology and the High Meadows Environmental Institute for their support in the final stages of this book project. I hope their faith in me is borne out. I am especially humbled by the graceful humor and administrative prowess of Mike Celia, Kathy Hackett, Rob Nixon, Serguei Oushakine, Carolyn Rouse, and Gabriel Vecchi. João Biehl, John Borneman, Lisa Davis, Julia Elyachar, Agustín Fuentes, Monica Huerta, Rena Lederman, Paul Nadal, Lawrence Ralph, and Yu Xie have been patient and giving. I look forward to all the things we have to learn together. And between Santa Cruz and Princeton, I've had the uncommon luck of being a member of two amazing junior faculty cohorts. Nidhi Mahajan, Tsim Schneider, and Savannah Shange; and Hanna Garth, Ryo Morimoto, and Lauren Coyle Rosen.

The book's main arc coalesced in an unruly Google Doc in furious write-chats with Nick Shapiro and Tim Choy. Corey Byrnes, Joe Klein,

Andrew Mathews, Anand Pandian, Lisa Rofel, and Mei Zhan workshopped a manuscript draft in Santa Cruz, just before COVID-19 lockdowns went into effect. Their inexplicable and entirely unearned generosity reinvigorated, in one afternoon, my faith in this strange vocation. This exhilarating day also cemented the final structure of the text. The momentum of their engagement carried me through, in exhausted inertia, to finishing the manuscript during lockdown. Hollianna Bryan gave extensive feedback on an early draft of the book and Aaron Wistar, Stefanie Graeter, Michael D'Arcy, Lisa Davis and the Fall 2020 Princeton Anthropology proseminar, and Micah's cohort at UBC Geography helped me think through it at the end. Nancy Gerth worked tirelessly on copy edits, and J. Naomi Linzer prepared the index valiantly. Three reviewers provided valuable feedback. They took the project seriously on its own terms, and indeed helped me clarify what those terms were for myself. The project is more itself because of their attentions. Kate Marshall and Enrique Ochoa-Kaup have been a wonderful editorial team at UC Press. They shepherded the project through its various stages with grace and understanding. Julie Guthman was an important advocate for it at the *Critical Environments* series.

The time and care of many remarkable scholars shaped this text. The PIAO Interdisciplinary Atmospheric Collective—Tim Choy, Nerea Calvillo, Dehlia Hannah, Nick Shapiro, and Jerome Whitington—made an atmospheric fellowship. Candis Callison, Vivian Choi, Madeleine Fairbairn, Fan Ke, Derick Fay, Jim Ferguson, Elizabeth Ferry, Mike Fortun, Kim Fortun, Anne-Lise François, Mette Halskov Hansen, Donna Haraway, Hong Wei, Andrew Lakoff, David S. Jones, Yoon Jung Lee, Sara Mameni, Joseph Masco, Jeffrey Moser, Micah Muscolino, Michelle Murphy, Hoon Song, Rei Terada, Chris Tong, Max Woodworth, Emily Yeh, and others read and commented on various sections in conference and other settings. I am indebted to many others for moral and intellectual support along the way: Chloe Ahmann, Andrea Ballestero, Franck Billé, Alex Blanchette, Zac Caple, Mun Young Cho, Jason Cons, Michael Eilenberg, Rune Flikke, Amitav Ghosh, Bridget Guarasci, Gokce Gunel, Ilana Halperin, Christine I. Ho, Colin Hoag, Matthew Kohrman, Ralph Litzinger, Zeynep Oguz, Juno Salazar Parreñas, Jake Silver, Kristen Simmons, Noah Tamarkin, Sarah Vaughn, Kaya Williams, and Amy Zhang. Gaston Gordillo, Michael

Hathaway, Shaylih Muelhmann, Ada Smailbegovič, Yana Stainova, and Derek Woods filled Vancouver with ideas and enchantment.

With love: Natasha, Huma, and Zavain Dar are family. Molly Cunningham has stood by through multiple reboots. Micah Hilt's support and care was constant. His photography is included throughout its text. Eric Tran enjoins me to remember the world-making magic of words. The kindness he has shown to my family in the final days of my writing is incredible. Vivian Choi's presence and friendship fueled my writing. Nick D'Avella is steady and kind, all water and stone. Mark Fleming reminds, with soft heart, *Be Excellent to Each Other! Party On, Dudes!* Stefanie Graeter and I share the trail. Michael D'Arcy is poetry and grace. Alissa Bernstein reminds me to be grateful and to be humble because it's always the best day of our life.

My family: Nancy Malancy Zee-Hilt cuddled through the world-cracking anxiety of the last year of writing, woof woof! My many aunts, uncles, and cousins are a net that always catches me. To my siblings Oliver, Kelly, Steven, and Bee Hui: you've grown on me. With hopes that we'll continue growing and learning together. Thank you to my parents, Tuen-Jiun and Renee, whom I love more than anything, and who have supported me through more than a decade of trying to get me to explain what anthropology is. Their doors and hearts have always been open for me. I have needed both.

Three of my four grandparents passed since I started the book. Their lives and migrations trace an arc across the Pacific, a transpacific terrain whose trails I have only just realized coincide with this book's wandering fieldsite. In sifting through dust, I find our family's diaspora ever again. Though they could not have read this book, I wish that they could have held it. They are the place where the wind will always blow back.

What follows is for them. It is a contribution toward unpayable debts, a coin tossed into the sea. My *aniang* and *aya*, Hwa-Chuo Yen and Yee-Fat Zee; and my *waigong*, Ching-Teh Chow. And for my *waipo* who survives them, Kay Chow, to whom I can finally answer: the book is done.

As this book enters publication, we mourn Sam Dubal. I hope that a bit of his enormous spirit glints through the words to follow. His loss puts me beyond myself.

I'm saying
    the dogwoods
cried themselves
    sterile and still
my friend is gone[1]

Summer wind dislodges the corner of the plastic tarp from a brick weighting down its corner. The sand underneath frees into the wind. The tarp flaps cacophonously, amplifying the steady roar of the night. With the rustle of the leaves of the poplar windbreaks that mark the edges of the village, and the sand pattering hard against the windows, it sounds like nothing less than a summer storm. Dry rain and crinkling, plastic thunder. I startle awake on the other side of the brick wall from Old Tai and his wife.[1] They are up now, too, and outside. Caught in the wind, sand streams out from under the tarp and fills the half-finished courtyard. In the night, a dust storm gathers in the walls of the courtyard of their government-issued house (fig. 1).

The Tais, in their fifties, have lived in this pilot relocation village now for five years. The desert's yellow streak, a voraciously expanding dust bowl, surrounds the village on three sides. Their house occupies a corner of the small grid of identically walled plots, a resettlement zone that was appended to an existing village to lure herding families off their pastures. Resettlement is part of the local dust control strategies that have poured the lives of families into the raging flow of sand and wind. They are ex-herders, too, evacuated here off degraded grassland. Old Tai often speaks

Figure 1. Tai family courtyard. Photo by author.

of their old life "in the sands" (*shazi li*) as a prior act in a drama of sand and state playing out on the family's sandy pasture. Soon after arriving, Old Tai converted the area meant as a shelter for a milk cow or a few sheep into a room for his elderly parents-in-law, who sleep among sleeping farm equipment and hard-baked *mo* hanging to dry from the rafters.

I sleep in the room they have built for their two grown sons. Both are long-haul truck drivers who have never lived in the resettlement village. They stay on the road for months at a time. They haul water, saplings, and barbed wire for the state forestry schemes rising everywhere to keep the earth on the ground of desertified pastures. Their family was dispersed by the sand, and now they disperse through it: the couple displaced by state antidesertification measures; their two sons trying to make it in the economy taking shape to hold sand, their trucks bought on credit, plying their way through the oceans of dunes.

In the daytime, Tai works with his brother-in-law and their new neighbors evacuated from *in the sands* into the resettlement village. Old Tai points toward the Tengger Desert on the hard, yellow horizon. He explains, "Sand from here is unsuitable for concrete." It is so fine that it rises on even moderate winds, hanging in clouds or moving as storms that have earned the Alxa Plateau the dubious distinction of "cradle of dust storms for Northern China." Surrounded by it, Tai nonetheless has to buy sand for construction. Sand, sand everywhere. Stirred into a slurry with wet cement, it is a crucial resource for the small construction projects that occupy the family on long summer days. Sealing the walls and floors of the Tais' state-issued and still state-owned house, they reckon, is not simply a matter of improving the structure but also a means of cementing their claim to it. Perhaps after the pilot period of the antidesertification relocation project, they can make a solid case that the house should devolve to them. Improving the structure is a security measure they deem necessary to stake their claim to the house into its very walls and foundations.[2]

Phases in their life align into phases of sand. Biographies chapter through sand's movements and changes. Its spread spreads them apart. Sand, bought and ready to mix into cement, might harden into a provisional permanence, steadying their house and claim to it against the whims of the state and the heavy, scraping wind. Their diligent conversion of sand into concrete makes their work to improve the structure a way of limning a confounding present, marking time until the land turns steady again. In the nightwind, we try our best not to breathe in their unfinished house.

In the crashing night, wind and sand catch each other in a dance of geophysical phases. The couple scrambles in the dark, clamoring to break the suspension, and in doing so, they are swept into its formation, their every movement shaped by, shaped into, tangling earth and air. Old Tai rushes to secure the tarp over the quickly dwindling pile, and I grab bricks to weight its corner. Aunty Tai shields her face with a scarf and grabs a wide broom to catch the stream of feral dust frothing out of the naked heel of the heap. She sweeps in a smooth arc to catch the flyaway grains and return them to the ground. Twisting against the flow, her body shapes itself against the low spiral of dusts. She turns like a bird whirling in the summer wind, "a small part of the pantomime" (Stevens 1954, l. 11). Her broom and body unfurl a counterspiral in the tiny storm, forming and breaking in the air around her.[3]

In their contention with the liberated heap of sand, they take part in an unfurling geometry of materials, living China's late socialism as their entry into a storm. They reach into surfaces of contact to wrest apart the turbulent touch of piled earth and raging air, two substances as they coalesce into a geo-atmospheric event: Aunty sweeping her broom in a cyclone against the cyclone; Old Tai and I working together to resecure the plastic tarp across the pile, a contact zone where sand and wind meet and where our intention rubs against the relentless potency of earth-moving air. Bodies in tempest, a storm of bodies. Our actions strew across the thresholds where matter shifts into matter, where the sand that might have become concrete walls instead catches into the thickness of a storm.

It may be tempting to imagine their efforts as a matter of exerting control over an environmental process, an inanimate world still in the turbulent waiting room of its own pacification. What might we make, instead, of their efforts, their calculations, their very bodies in this choreography of substances: sand and wind, unbuilt house and dervish broom? Pulverized land, trucked in and purchased, only to ride inhalations that draw bodies into the storm and the storm into bodies? The state? They have become part of a geophysical choreography. Our bodies and actions for a moment take on the shape of the dusts they seek to control. Aunty's parents, woken by shouts in the rainless tempest, watch from their unlit window.

Introduction

EARTHLY INTERPHASES

In early April of 2001, two closely spaced pulses of Siberian jet wind blew into semiarid Inner Mongolia, whipping surface sands into the air. The western reaches of the region, at the frontier with Mongolia, were grass-bare. The convergence of years of drought, a mounting crisis of land degradation, and an unusually warm winter left vast tracts of the region's sands exposed and unstructured in the early spring. In these conditions, an early thaw of sandy landscapes aligned with the onset of strong seasonal winds.[1] An inbound temperate cyclone system scraped against the loosened earth, peeling the land from the ground as a dusty emission, and moving it as a rapidly evolving weather system. The two pulses of wind converged into a complex of airborne dust that rushed toward Beijing. They swirled, over the next weeks, into a single storm whose geophysics and geochemistry would move along its planetary course.[2]

Over the course of the month, the cyclone of land surged eastward. It filled seasonal airstreams in bursts of earthly color. Its yellow rivulets moved in a complex trajectory of smooth lines and crinkling twists.[3] As the storm moved across northern China, coal smoke, soot, volatile organic compounds, and the industrial effluents of China's booming economy entrained into its mix, glomming onto particles of aerosolized land,

Figure 2. "A Perfect (Dust) Storm," with high-resolution inset. April 7, 2001. Courtesy of NASA Earth Observatory.

altering the geochemistry of the storm. The storm quickly breached the dotted line of the Great Wall and, two days after forming in Inner Mongolia, it fell over Beijing as a bout of dust weather, one of that season's eleven major and nearly consecutively spaced major dust events.[4] On the threshold of a long-awaited "Chinese Century" (Pieke 2014), the storm, its noxious particulate density, was the country's interior passing over itself in its surge toward the dust-battered capital: modern weather.

NASA reported on the storm event as it swept over Northern China two days after its initial formation in the interior of the Asian continent (fig. 2), juxtaposing satellite photography with visceral testimonials from people swept into its path. At the surface, the density of the storm confused day and night. Ground-level eyewitnesses "reported that around 7 a.m. local time" in the Chinese Northeast, "the dust blocked enough sunlight to leave the skies as dark as midnight, and reduced visibility to roughly 20 meters." They described their view into the dust as an occlusion of vision. Perspective, fixed inside the storm, is a mode of unseeing, an optics preempted in the eclipse of the air by the sky.

In the impossible freeze frame of extraterrestrial image, however, the storm blazes reflected light into space, revealing itself to the machine eye of NASA's Earth Observing System, its gaze fixed on China from low earth orbit. From above, the muddle of confused time and stunted vision at surface level furrows into the sharp, crisp lines of an earthform stilled by the camera. Satellites photographed the storm as an airborne landscape, complete with its own shifting geomorphology, held still in the image as a relief map of airborne valleys and rippling dune formations.[5] In its pedagogical interpretation of the image, NASA enjoins the reader to attune to the storm as both landscape and meteorology. The storm, it offers, "almost forms its own topography, with ridges of dust rising up below the clouds" (NASA Earth Observatory 2007).

> Ridges of dust
>   rising up below
>   the clouds

The god's eye squints to make sense of the disconcerting clarity of the picture. The China above doubles and obscures the one below, still visible at the storm's fraying edges and through the intermittent skylights opening where dust thins. The desert below has phased, doubling itself into a desert above: land whipped into floating land, a weather event now creating its own weather.

The sky was full.[6] There were mountains in it. Lofted into the atmospheric foreground, the storm is a continent in dust, rising and falling with the surge of the spring winds. Pictured in multiple simultaneous phases of a geo-meteorological process, *China* is both the territory and its uncanny meteorological double, and the shifts in terrestrial phase between them. This China rises and falls. Its earth is a plume, unfurling with the coming of spring. The continent had become a constituent of "sky of our manufacture" (Taylor 2016), unfolding the brittle line between Nature and Culture into a vertiginous interface of socionatural entanglements.

———

Continent in Dust is a political anthropology of strange weather. It is an ethnography of what I call China's meteorological contemporary—the transformed weather patterns whose formation and fallouts have accompanied decades of breakneck economic development.[7] As the headiest days of rapid economic development settle into difficulty breathing, Reform and Opening, at various points along the course of storms, has also opened into questions of how to persist, adapt, and suffer through bad air. The book inquires into Reform and Opening as an array of political, social, conceptual, and technoscientific experiments. Each of these experiments grapples with the curious propensities of modern land and air to phase into one another, and in doing so, raises profound and practical questions for politics, bodies, and analysis. We approach each of these experiments as they offer ways of attending to the beginning of the twenty-first century, and the fourth decade of Reform and Opening, in China and downwind, as a condition of meteorological emergence.[8] In dust, terrestriality and meteorology evince one another in a profusion of phases, an elemental choreography that unfurls possible Chinas.[9]

The weather had changed. In the thick of China's geopolitical ascent, dust storms substantiated the capital's air as a consequential suspension. As 2001's season of dust storms was beginning to settle, Beijing exploded in wild speculation over the causes of and potential resolutions for this dangerous mineral weather. Planners openly fretted over the expanses of mobile desert sands lurking and lurching at the threshold of the capital. The possibility of the burial of the capital in mobile sand, by advancing dunes or particulate matter unloading from the sky, was openly discussed in official circles and state media. Weather events and aerosols in particular—a mounting crisis in particulate air pollution and catastrophic seasonal dust storms that was quickly becoming a signature of Chinese cities—appeared as shadow-histories of the present, a meteorological aspect of a time most often narrated through rapid development. In "the first decade of the new millennium," dust storms "evolved into one of the most widely and controversially debated environmental issues in the People's Republic of China" (Stein 2015, 321).

This book traces out this explosion of dust into Chinese politics as a conundrum of how the political dynamics of Reform and Opening interact with aerosols: mixtures of particles and airs, earths and skies, that form, drift, and break along the course of the wind. In our inquiry, particulate dynamics appear as the fallout of explosive economic growth, and as a material condition that gives traction to unexpected configurations of relating, breathing, and governing in the twenty-first century. In each of its scenes, the official histories of development and national arrival are offset into the geophysics of the aerosol and meteorological phenomena that have apparently accompanied development. What shadow-histories of the future might be possible at the near miss of two material histories of China? How can we hold *both* Chinas in view—the satellite's China in the sky and the one that it obscures below? As an ethnographer in China's meteorological contemporary, I ask: what if the rise of China were to be approached literally, through the rise of China into the air?

The confluence of meteorological derangement and meteoric economic growth raises the question of Reform as a time of strange weather. Sudden infusions of particulate matter into the capital's airspace in the early years of the twenty-first century anticipated explosive debates in China's cities and social media over PM2.5 a decade later. Thick hazes of dust, soot, and exhaust cloud the muscular central messaging that Beijing had finally returned to prominence on the world scene, a proclamation rendered unstable in the changing colors of the sky. Beijing's bid for the 2008 Olympics was submitted in the immediate aftermath of the worst dust storm season in China's modern history (Jeux olympiques d'été 2010, 21–22). The sunny image of a Green Beijing Olympics was premised upon the notion that China's ascendence could be evaluated by its ability to control the particulate matter in the city's air for the benefit of foreign spectators and world-class athletes operating at below peak performance by virtue of breathing Chinese air. Against the spectacle of incoming drifts of mineral dust, the management of the air, its contents and its dynamics, would become a crucial proving ground for evaluating the capacities of the modern Chinese state.

By the turn of the twenty-first century, dust had become a durable feature of the northeast Asian springtime, reliably reported in weather reports across the region. For countries downwind, dust had quickly become a matter of incoming drifts of foreign land. A day after passing over northern China, major dust events leave Chinese airspace, but not before accruing atmospheric effluents into their suspension from industry, power plants, and other sources before passing over the Korean Peninsula and then Japan. Worries over thickening political and economic ties with a rising China sublimate into vocabularies that inflect geopolitical anxieties into words for bad weather: *hwangsa* in the Koreas, and *kosa* in Japan: "yellow dust," for the telltale hue of a foreign desert (Kar and Takeuchi 2004). Under a strong enough wind, the finest particles of desert can remain suspended indefinitely, engrained into the geochemistry of the troposphere: a becoming-Chinese of planetary atmosphere.

We begin in the dusty middle of this weather system, tracking aerosols like dust and particulate matter as they signal collapse and also condition new political and environmental possibilities. We focus especially on the dust storms and particulate matter events that have transformed the texture of both political governance and everyday life. Our attention condenses, floats, and scatters along dust-transporting airstreams, lingering with people at various points in the trajectory of a storm, for whom "China" exists in the potent interphasings of land into the geophysical substrates of aerosol weather systems. Aerosol transitions, movements, and scales draw the ethnographer and his interlocutors into "a field of accidental social relations" (Rosaldo 2014, 108): dust, that is, is not only the object of a shared fascination, but the very medium through which relations between people and between institutions take shape. Scientists and engineers, officials and herders, breathers, artists, and anthropologists encounter one another through choreographies of dust.

In my fieldwork, dust was most often described to me through reference to the shapeshifting and relational materiality of the substance that aeolian physicists call *fengsha*, or *wind-sand*.[10] For this reason, my inquiry into the worlds and planets that open with dust stays close to *wind-sand, and* its curious materiality of transitions, as a guide. *Wind-sand*, following R. A. Bagnold's description of its closest English cognate "blown sand," reveals the planet through phase shifts. *Wind-sand* is not reducible to its component parts, as its properties cannot be derived from wind nor sand in isolation. It is instead "a new kind of flowing substance" born of their specific relating: sometimes a field of mobile dunes, sometimes lung-penetrating particulate suspension, sometimes hemispheric dust event. To trace *wind-sand* is to be captivated by questions of the many formats that their substantial relation can take (Bagnold [1941] 2005, 105). Staying close to *wind-sand* means that political analysis cannot be confined to the mechanical evaluation of the successes and failures of state programs aimed at fixing the planet against its change. Instead, fantasies of control open into other geometries of agency and inertia, just as the Tais and their anthropologist, in their attempt to contain their pile of sand, are formatted in its flow, their agency distributed into the whirl of substances.

Dusts emerge as multiple in the kaleidoscopic capacities of *wind-sand*. It is an entailment of its unfolding and processual materiality (Zee 2020e); just as the perturbation of a kaleidoscope-turn unsettles new patterns into being, we attend to the moments of apparent breakdown that dusts often signal as moments in which other configurations come into being.[11] And so, we approach desert, dust, and storm as possible phases of *wind-sand*: as multiple possible permutations of a single processual materiality. *Wind-sand*, for instance, is in play when herders resettled into villages describe the abrasion of dusts that scrape grass off a windy pasture or scrape exposed skin or hide. It is also what environmental engineers consider when they picture a dune in potential motion or painstakingly maintain infrastructures to hold drifts of dust below the thresholds of suspension. *Wind-sand* is a relational substance that clasps meteorological and geophysical conditions as part of the same complex process of transition. And it is also the rubric through which open-ended relations of air and earth can be apprehended as an array of possible arrangements, phases, and dispositions.

If, following Mary Douglas, dirt is matter out of place, *wind-sand* might be matter out of phase: "a cross-section in a process of change" (Douglas 2005, 39) that relates sand, dust, and storm as patterns of one another. Theorists of dust have attended to its elusive materiality. "Dust is not just 'matter,'" in any stable way, but it is "something that troubles our notions of matter" (Parikka 2015, 88). Dust is what is shed (Marder 2016, 5), the unbearably light and dense physical manifestation of political formations that operate through ecological ruination, and so it can be read as the trace of destruction. But *wind-sand* and its permutations of geo-meteorological substances unsettle the sense that any arrangement can be described with the finality of an ending. *Wind-sand* reveals, again and again, a planet made not in fixed forms but in the phase shifts between them.[12] Kicked into sky, China suspends, a meteorological stratigraphy in particulate matter.[13]

———

The book addresses the following questions:

First, it asks how the problem of materiality in ethnography can be brought to bear on questions of governance, politics, and the state. What I am after here is something more than the observation that materiality matters, or that materiality itself undermines given constellations of the political by exposing their anthropocentric limitations. This kind of argument tends to reproduce an agonistic and binary account of politics and environment, one that implicitly identifies environmental processes with resistance against power. I seek to cultivate an art of noticing that begins with noting the limitations of existing political imaginaries in order to attune to how they are reconfigured, entrained, and patterned into the meteorological dynamics that they seek to control. And so, my attention is trained on how political formations might shift into other configurations as they enter the choreography of the world in all its vibrant materiality (Bennett 2010). If *wind-sand* charts out a course of planetary emergence through its phase shifts, how might it also induce reconfigurations in the architectures of institutions, bodies, and relations of all kinds?[14] As *wind-sand* unfolds into itself through a cascade of geo-atmospheric permutations, so too do the determinants of an anthropocentric politics reconfigure through more-than-human experiments.

A second question: how can the dynamism and open-endedness of weather systems in *wind-sand* stoke open-ended transformative modes of governance? Karen Barad writes, "The world is an open process of mattering through which matter itself acquires meaning and form through the realization of differential agential possibilities" (2007, 141). If we refuse to categorically contrast this open process of mattering with the rigidity of anthropocentric politics, we must attend to the political not in terms of rigid logics, but in terms of experiments that grapple, in real time, with the planet as it changes. I return to the question of experiment as a way of contributing to an "anthropology of becoming," grounding my understanding of the political in "the intricate problematics of how to live alongside, through, and despite the profoundly constraining effects of social, structural, and material forces, which are themselves plastic" (Biehl and Locke 2017, x). Experiment is a notion through which a history of modern Chinese politics can be traced. Through it, I insist on the open-endedness of political and environmental formations alike, becoming with one another in conditions of modern weather.

The experiments that concern this book depart from many of the functionalist assumptions of ossified liberal *and* socialist political traditions. Lisa Rofel, in her ethnographies of Reform and Opening, emphasizes "the nondeterministic content and direction of the reforms" that, in retrospect seem planned. Her account traces out scenes of contemporary life and power in China, eschewing the notion that it can be interpreted as the straightforward implementation of a rigid political logic or ideological orthodoxy. Instead, Rofel centers scenes of encounter, for an ethnography that "challenge[s] that ontology of pure categories" (Faier and Rofel 2014, 373). She offers a counterpoint especially the typological impulse that seeks to purify a coherent theoretical model of "Chinese rule" out of what can only be understood as an array of situated encounters and experiments. Departing from studies that pose "environmental challenges" in China only insofar as they facilitate the perfection or fragmentation of an already assumed mode of authoritarian rule (Mertha 2009), I propose that tracking late socialism as an experimental formation offers a view of Chinese environmental politics that does not simply follow out a fixed plan, as though already "fully laid out, based on normative principles" (Rofel 2007, 8).

This leads us to a third question. How might ethnography be reconfigured and extended to attend to questions of planetary and political emergence where they crosshatch with one another? The embroilment and co-constitution of political and meteorological formations makes evident, following Mei Zhan, a pressing "need to co-imagine a critical methodology oriented toward continuous unfolding and differential becoming" (Zhan 2019 187). With *wind-sand,* anthropology may come "unmoored from its classical objects and referents" (Jobson 2020, 261). The modes of existence of both *wind-sand* and ethnography are configurational. Each requires expansive accounts of relation and becoming, and each demands an attunement to how our senses of what is and can be assembled into other shapes. In this welcome unmooring, anthropologists find space to sift and shift through the various traditions and affordances of ethnography itself, "identifying and expanding the scope for what remains on the threshold of possibility" (Pandian 2019, 4).

Implicit in these questions is the idea that power must not be understood as a one-way action onto environment, as if political formations were external to the geophysics of their Earths. The histories of Reform cannot be abstracted out of the geochemical content and density of the sky. Worlds cannot be excised from winds any more than the wind itself can be stopped. Dreams of geophysical stabilization crack open. Designs to reorient relations with environments are revealed again and again to be part of the complex dynamics through which environments emerge. The significance of these strategies and maneuvers thus exceeds any straightforward accounting of the "environmental impact" of China's rise and its impacts for climate change, to be debated in international meetings, where countries are straitjacketed into the technocratic language of "emitters." Rather, I demand attention to planetarity itself as a site of political experiment, opportunity, and contention.

As dust shifts from a problem to be resolved to a condition of planetary emergence that drives Chinese officials to pose new relations between the longevity of the Chinese state and the sustaining of the planet, the meaning of China as a planetary and political question is posed and posed again. Across experiments and weather systems, the question of China is continuously reassembled, appropriated, and retrofitted into the technopolitical dynamics of the earth system "in a moment when the latter has become a technical object" (Woods 2019, 9), and also a demand to imagine more robust and more livable figures of relation on a changing planet. We thus explore, in what follows, the strategic and experimental procedures that seize and remake regional, global, and planetary scales—especially those that figure Asia as a laboratory of possible planets. These make earthly connections through dust and its choreography across earthly and atmospheric phases. None of them default to the planet or the environment as synonyms for "everywhere."

With this in mind, we explore the weather systems of late socialism through the topologies of power (Collier 2009) that take shape through it, warping the wind and the state into each other's shape.[15] The disorientations of *wind-sand* guide us through the reimagination of the political as it decomposes and repatterns in the flux of bad weather. They allow us to understand both political experiment and planetary emergence as questions of phase shift. And they require an ethnography that takes part in this "continual unfolding and differential becoming" (Zhan 2019, 187). Environmental change, political experimentation, and the ethnography that transforms with them are thus each part of the recombinatory physics that we have called weather.

MODERN WEATHER

The year 2000 was, according to the Chinese Meteorological Agency, the most severe documented dust storm season in the half decade since the founding of the People's Republic of China in 1949.[16] Its nine major dust events set records in both the frequency and intensity of dust storms pummeling Beijing. This record was only to be surpassed the next year, in 2001, with its eleven storms. In 2000, in the face of a gathering political storm in the dust-shocked capital, then premier of the Chinese Communist Party Zhu Rongji left Beijing with a coterie of more than 350 party officials. Zhu and his entourage traced the paths of dust storms, beginning in Beijing and running aground in their purported sites of origin, following the wind in reverse as a movable segment of the state.

Their journey traced an airstream backward out of Beijing into the Chinese interior, a route revealed by the meteorological mapping technique of back trajectory modeling (see Chen et al. 2017), which reconstructs weather systems backward in time and space from a designated point. In doing so, they cross-hatched administrative and meteorological geographies. Center and periphery and upwind and downwind relations tangled in the movement of the premier. Distinct cartographies took on each other's shape as the junket charted a weather map of environmental insecurity rumbling through spaces of uneven state power.

In train with the premier, high government officials parsed the territory for its aerodynamic qualities. They posed far-flung landscapes as chokepoints in an unruly modern weather system, its variegated geography of upwinded dangers still coming into view. The universe of possible government interventions patterned out with the fraught meteorological prospects that they pictured across the national territory. The junket moved to Hebei province, just past Beijing's administrative boundary, where mobile sands blew against farmhouse walls, making ramps for goats onto roofs. They streamed against the wind to Gansu Province, where, in his early career as Party geologist, Premier Zhu had warned against desertification.[17]

The caravan paused at the Alxa Plateau of western Inner Mongolia, hundreds of miles and two days by dustflight from Beijing. At the desert's edge, he invariably characterized mobile sand and new deserts as a crisis of earth surging toward the capital. The media event that crystallized around the premier's airstream tour anticipated, in a tableau of sandy shapes, the coming burial of a Beijing preparing feverishly for its debut into the rarefied echelon of world cities. The relation between upwind land and downwind weather was, in Zhu's warnings about *wind-sand*, coming to formal coherence as a condition for a becoming-meteorological of state power: dust storms were literally land, and land, he warned, was on the move.[18]

The assurance that that "land stays in place"—among the "core elements to land's material quality" (Li 2014, 591; see J. Klein 2019)—was frustrated by its insistent entanglement with the wind. For Zhu, land was quickly becoming a question of volumes, speeds, and flows. It proceeded in the devouring advance of sand, mobilized toward the capital by spring winds. Land was *both* political territory and also a theater of earthly interphasings. One could not say *sand* without hearing *wind*. And in this, China's land, in *wind-sand*, became a threat to a capital whose long history has been marked by anxiety over invading forces.[19] It lofted on the jet winds that would, with each spring, redistribute the country's sandy interior as stifling weather in the very centers of state power.

The Alxa Plateau, an out-of-the-way place (Tsing 1993),[20] had become evident from downwind as an official "cradle of dust storms" in Beijing's dust-shed. Its land degradation—a regional economic and ecological crisis in the heavily pastoral region—was insinuated into the "northwestern route" of dust emissions and transport, traced backwards from the airspaces of insecure, arriviste power. With the desert at his back, Premier Zhu spoke as if at the headwaters of a river of flowing, floating land. This farthest reach of a national wind had become a threshold in an aerosol process. Its history and ecology, its people and its local government, were suddenly cast as the conditions of an environmental machinery gone haywire, to the detriment of more powerful places on a shared wind. Pictured from the distant capital, Alxa had become a danger zone in a new national geophysics. Facing his gathered entourage, Zhu proclaimed that if its deserts could not be controlled, "sooner or later China will have to move its capital" (*Renminbao* 2000).

State power grapples with meteorological emergence. Zhu's migratory junket traced the course of the dust upwind to the aerosol threshold of the Alxa Plateau so as to warn of the displacement of Beijing itself by the bad weather unleashed in (or by?) Reform and Opening. The speech is a cautionary tale of contending political and environmental forces, one in which the political geography of a prosperous and ascendant China faces its cartographic undoing by storm and by sand. A curious political meteorology was taking shape, born in dust and the insecurity that gripped the capital, now a beleaguered island in the path of a torrent of broken, windswept land.[21]

China's meteorological contemporary, as it is presented in this book, is "less a theoretical framework than a mode of attention I wish to sustain" (Choy 2011, 11) through attending to specific embroilments of *windsand* and power. Its juxtaposition of adjective and noun is not a seamless merger, but a craggy and uneasy terrain of partial connections. Where they meet there is no emerging certainty: an "answer is another question, a connection a gap, a similarity a difference, and vice versa" (Strathern 1991, xxv). China's meteorological contemporary is also a way of making sense of what I describe below as late socialism, an experimental modality rather than a claim on the now as an ontological rupture in political or environmental time. Neither term—*China's meteorological contemporary* nor *late socialism*—is, in this sense, a periodizing claim that distinguishes a now from what came before. Both are exhortations to attend to the many weird little shapes that time, weather, and China bend into in specific experimental formations.

By some measures, northern China has experienced a nearly fiftyfold increase in storm events in the fifty years since the founding of the People's

Republic in 1949 (Liu and Diamond 2005, 1183). This rapid increase in dust events is closely correlated with the processes of land degradation officially denoted as desertification (*shamohua*), sand-ification (*shahua*), or wasteland-ification (*huangmohua*).[22] The incidence of both desertification and dust events has accelerated even more in the decades since the late 1970s, the period officially designated as China's era of Reform and Opening. Environmental degradation, especially in semiarid pastoral regions has long been blamed on the corruption of local cadres, the mismanagement of pre-Communist Chinese regimes, and, most routinely, on pastoralists themselves, setting off new state programs for managing ethnically Mongolian herders and pastoral landscapes in the 1980s and '90s (Williams 2002). Dust events nearly doubled in the 1990s as desertification, today, afflicts more than a quarter of Chinese land (State Forestry Administration 2011).

These geo-meteorological transitions have been contemporaneous with Reform and Opening as the explosive economic growth afforded by socialism with Chinese characteristics accelerated to its most breakneck pace (X. Wang et al. 2004). Dust storms have thus appeared prominently in accounts of China's recent environmental challenges. For observers in and outside of China, dust storms and airborne particulate matter invite speculation on the longevity of the state. In historical research, dust has often stood as evidence of the unraveling power and legitimacy of the current Chinese state, and the climatology of dust events serves as a convenient environmental correlate on the waxing and waning of dynasties (Chen et al. 2020). Much reflection on the Reform era has it narrated as a time of environmental crisis on a scale that exceeds economic growth, explicitly posing the environment as such as an existential threat to the Chinese Communist Party and its political status quo (Economy 2005). The state either threatens to collapse or hardens into an ever more fixed authoritarian stance. Environmental trouble in China has alternately stood as proof, across a wide spectrum of political and analytic positions, of the depredations of the Chinese embrace of capitalism against the alter-futures of a socialism that lives, cynically, in name only; or of another kind of cynicism, deeply embedded in Party rule; or, smugly, of the self-devouring growth of an upstart rising power (Livingston 2019).

National ascent is dizzying. Ethnographers of late socialist China have sought to capture "the vertiginous sense of imbalance that the rapid pace of change induces in those most caught up in its embrace" (Anagnost 1997, 15). The experience of national time in the Reform era is a phantasmagoria of change, replete with exhilarating confusions and new subject positions (Rofel 2007, X. Liu 2000). Marshall Berman calls this heady mix of pleasure and danger *modernity*: a "mode of vital experience—experience of space and time, of the self and others, of life's possibilities and perils" (1988, 15) that explodes the smooth arrow of developmental time into a riot of fraught and irresistible affects. Berman diagnoses modernity, felicitously, in a near-meteorological metaphor. At once anxious and hopeful, terrifying and exhilarating, modernity is the experience of being poured "into a maelstrom of perpetual disintegration and renewal" (Berman 1988, 15).

Dust storms and national arrival evoke each other, a slant rhyme of disoriented worldings. Each is a version of a country roiled upward, crosshatching over Reform as an entrainment into the erosion and pull of multiple whirlwinds. In a dust event, the earth whips into atmospheric suspension; Berman evokes Marx to describe this breathless jouissance of change, consonant with descriptions of China's other modernity: "all that is solid melts into air" (1988, 15; see Rofel 1999). Melting, I must emphasize, is not disappearance or obliteration. Melting is a shift in phase, a reconfiguration of matter. It is one possible way in which one arrangement may become another.

The melt of all that is solid into air is a phase shift. It is a movement in a dance of substances, processes, and relations. A storm does not negate the land but rather confirms its capacity to change. This phase shift poses the "remarkable figuration of solidity" (Choy 2011, 144) that girds social thought against its fear of the wind as a scene of possible emergence.[23] And so, modernity might also be a question of making sense of what Tim Choy calls "air's substantiations," how air comes to matter "in deeply felt, variegated, and variegating ways" (2011, 143). Modernity, to think in the parallax between Berman and Choy's stormy metaphors, is a pattern that emerges in the turbulent convergence of disparate and yet intimately bound maelstroms. If, between modernity and modern weather, all that is solid melts into air, what is to be made of that difficult atmosphere and the worlds shaped in and against its thrall?[24]

Aerosols displace the immediately available ways of marking political and environmental time in and through China: *wind-sand* is at once *both* a material signal of the environmental ravages of Reform *and* the substantial process through which ecological breakdown shifts into the emergence of new political and environmental formations.[25] As it comes to matter in multiple places and projects, *wind-sand* stages the possibility of recomposing both official accounts of Reform as a time of ever-improving material life in China and fretted environmentalist accounts of China at the center of a widening circle of planetary destruction.[26] Entrained into the traction of sand and state, China's late socialism becomes a terrain of experiments in environment and politics. It, too, is accessible through phase shifts: change as a matter of reconfiguration, of matter and world and the geometries of power folded and folded again into new shapes.

Displacing the "charismatic mega-category" of the Anthropocene (Reddy 2014), China's meteorological contemporary folds these soaring metanarratives into the physics of particulate matter and air. *Wind-sand* snags the momentum of these metanarratives into "new swirls and eddies" (Robinson 2013), shifting the fatalism or pat melancholia of collapsing environmental futures into an array of other, stranger patterns. We must then eschew an environmental politics that runs aground in the description of political and ecological breakdowns. Falling apart, of course, is also a way in which other worlds come into shape, actualized in moments of phase shift that ripple through political and planetary formations. "Decomposition . . . is different than decay" (Pine 2019, 8).[27] The Anthropocene, is one way, among others, of attuning to the way that more-than-human assemblages work. It cannot be a question of how human life causes or governs environmental change, but rather, an attunement to how human life is continuously decentered and reconfigured in its rapport with a planet thrown past all thresholds.

The vision of *wind-sand* and political formations, already rapt in parallel, entangling processes of phase shift, offers a way of disorienting conventional stories of ecological transformation and teleologies of planet death. Collapses and breakdowns are, indeed, precisely where phase shifts might happen. When massive land degradation frees sand to storms, for example, officials might prophylactically incite some phase shifts in order to blunt or preempt others. As François Jullien argues in his discussion of *shi* (disposition) in classical Chinese thought, any given arrangement appears not as static integration, but as a vibrating interplay of dynamic tensions, populated with the universe of possible arrangements latent in the one at hand. "Within the actualized static form, a dimension of perpetual soaring flight" (1999, 78) is discerned—when a thing's actuality cannot but allude to its other possible permutations.

If "it is impossible now to pretend that modernity has resulted in human rationalization and control over nature" (Whitington 2019, 221), "control" must be revised into a mode of attending to human intervention as an uneven participation in planetary choreographies of unfolding matter. The geometry of the political and the environmental cannot be evaluated, then, in an agonistic drama of opposing forces, but rather, in the permutational dynamics through which planets and worlds are reconfigured with one another. An ethnography of this permutational question of relations and emergences would require a refusal of any the sense of fixity that would preclude "any transfiguring encounter from taking place" (Stevenson 2014, 2).

Through *wind-sand*, political transformation must be understood as part and parcel of the unfurling of an "atmospheric condition rather than the expansion of anthropogenic powers" (Choy and Zee 2015, 211). *Wind-sand* moves analytics and attentions across what Frédéric Neyrat calls a dis-archic flux (2018, 3), in which the smooth, closed loops of modernist ecology and physics wobble open into patterns and happenings whose ends remain unpredictable. Political formations, even those quickly written off as authoritarian, are as open-ended and dynamic as the weather systems that occasion their change. The interface of political and environmental change must then be approached as an ensemble of experimental zones and materialities. Here, the political—not to mention our vocabularies for it—reconfigures itself in the wind. The social, the political, and the environmental in this book, are not self-contained and *sui generis* domains, but rather a set of unstable and ever-changing domains whose relations make up the weather system of late socialism.

———

Meteorology, as a science, insists on accounting for the complexity of what makes and moves the atmosphere. It is, in this sense, a science of relation, one that thinks across vast scales as well as across actants that may remain separate in conventional humanistic and social science work. This vision of meteorology poses the question of how the domains usually studied by the human sciences come to matter for the atmosphere.[28] What is at stake in my invocation of meteorology is not a shift in attention from earth to air, but to the reconfigurations of matter and politics that makes geology and meteorology mutual phases in a singular process of planetary transformation.[29] Aeolian physicists displaced my understanding of land and air, and jolted me into understanding deserts and skies as permutations of one another as iterations of *wind-sand*. Thinking alongside their understanding of *wind-sand* allows me to hold dunes, aerosols, particulate matter, and other formats of environmental materials as multiple phases of a meteorological phenomenon. And so, even as sands are imagined to come from distant upwind countrysides and "pollution" is often posed as a question of toxics from urban industry, here, I attend to them together as embroilments of politics, particulates, and environment under the sign of weather.

I insist on describing this condition as weather—and not just as pollution—to highlight the increasing untenability of identifying the weather with an externalized Nature. Weather manipulation dates back a century or more and the modes of state environmental intervention that we will track in practice have little interest in the Romantic distinction between what is manmade and what is not.[30] In their work, for instance, environmental engineers enact human life and the traditional anthropological domains of sociality and economy as part of the complex environmental mechanics that govern the mobility of landscapes and the formation of storms. They approach human and nonhuman things as heterogeneous elements in a socionatural assemblage variously called environment, ecology, economy, or society, with important consequences for the shape of politics.

The sheer banality of the idea of weather is helpful. Tim Choy, in his discussion of "breathers," appreciates the term's conceptual vacuousness

(2011, 146), a blandness which invites modes of discernment and attunement unmoored from an already-fixed schema of explanation.[31] This vacuousness allows me to sidestep the familiar tropes of responsibility, causation, and regulation that the more loaded term *pollution* foregrounds. What we think of as pollution, with its implicit technical and moral call for the attribution of anthropogenic responsibility, defaults to a familiar matrix of agency and blame. Certainly, being able to ascertain blame in the familiar format of "polluters" is fundamentally important to any contemporary environmental politics or imagination of environmental justice. But *pollution* sets us off searching for causes, purifying responsibility out of complex histories and relations when discreet actors must be named. *Pollution*, because it figures the act of contamination as an event with clear perpetrators, ironically also enforces a conceptual separation between nature and human worlds, rather than to the open-ended, underdetermined phase shifts that I am describing as the weather.[32] Figuring pollution as an externality to an otherwise pristine environment tends to render the atmosphere into a grid of clearly demarcated responsibilities, rather than an existential and transformative medium whose political affordances and possibilities cannot be predicted. This shoehorns the transformative and disorienting powers of modern weather into the siren comforts of a grid of a politics that already cannot hold.

The weather, as a less determined map of cause and culpability, can instead be approached as a question of the experiments in politics and environment that take shape with aerosol formations. It insists on the open-endedness of such formations, rather than demanding that we fix their dynamism into a ready-made grid of responsibilities. It also insists on the deeply generative capacities of what is often glossed as pollution. Aerosols are both disastrous anomalies *and* the existential medium in which earthly life takes shape, as the meticulous attunement to the microdynamics of particle droplets in COVID-19 has shown. The weatherworld, the "zone of admixture and interchange between the more and less solid substances of the earth and the volatile medium of air" (Ingold 2010, S122), is thus a zone through which we can parse the history and politics in aerosol physics, atmospheric chemistry and toxicology.[33] There is to be no restorative return to a purified nature. History, political economy, and the entirety of human life, then, are not simply events of pollution, but part of the ongoing substantiation of planetary atmosphere, which is banal and extraordinary all at once.

If dust storms signal ecological disarray in the inland headwaters of deranged weather systems, they also demand an accounting of worlds and planets that emerge with their flow and their passage between phases of matter.[34] Anthropologists have turned to atmosphere to "deterrestrialize thought" (Howe 2015). What could it mean for an anthropology to approach atmosphere through dust storms, wherein the air is itself a possible iteration of land? In which the terrestrial is the substrate of solid, fluid, and aerosol earths? This would not require a rejection of the terrestrial as a confining earthly tether. Instead, it would develop through a curiosity toward all the shapes that land can take when it is uncoupled from the physical and conceptual strangleholds of merely solid ground. As Hugh Raffles reminds, "Even the most solid, ancient, and elemental materials are as lively, capricious, willful, and indifferent as time itself" (2020, 6). Tracing dusts, anthropology might disorient its anthropocentric charge, en-earthing itself with this geology of transitions and flowing, falling landmasses. It might displace itself with the "stony maelstrom of the stony blast" (Raffles 2013), unsettled into analytic and disciplinary permutations that track with earth and sky as moments in a continuum of earthly phases.

LATE SOCIALISM, OR A SINOCENE EXPERIMENT

I locate late socialism in the experimental systems through which the state and other political actors encounter and change with environmental materiality.[35] The term *late socialism* should not be understood as a definitive periodization. Rather, it is a way of marking a specific dynamic of continuity and change, one that defies the transitionologies and definitive endings of socialism that a term like *postsocialist* implies. Late socialism, in my account, is not in the coherence or longevity of any ideological program or specific governmental apparatus in China, but in an experimental ethos and method that are common to multiple formations of the state in modern China. It is a propensity to morphological flexibility in governance and politics, a rejoinder to visions of authoritarianism as a brittle and rigid political formation that will eventually crack under the weight of reality. My account of late socialism thus emphasizes the capacity to relational mutation.

This demands a mode of inquiry that is not invested in the discovery of a singular political logic that would offer a unified explanation for disparate political phenomena. Rather, it traces out a robust multiplicity of practical experiments—not only in governance but, deeper, in how political fields are constituted and how they transform.[36] This mutability, which continually generates and displaces the subjects and objects of state and popular politics, is the very condition of "authoritarianism" as a principle of situated transformation, and not the proof of its fragmentation.[37] Fixating on fragmentation fails to register the historical and continuing dynamism of China's idiosyncratic—and, I stress, continuing—socialism.[38]

I trace late socialism, therefore, as itself an experimental system, operating in an evolving set of parameters, and yet irreducible to any pregiven normative form of the political. Experimental systems, Hans-Jörg Rheinberger explains, are "tinkered arrangements which are not designed for merely repetitive operations but for the continuous re-emergence of unexpected events." Experimental processes are oriented toward apprehending difference and changing with it. They are "inherently open and unfinished systems" (1995, 112) that operate through continually unearthing possibility by exposing their own limitations. They are both embedded in histories and habits of knowing and inherently engaged in the technopolitical processes that undermined those habits and reformat them to other possibilities.[39]

This account of experimental systems—open and adaptive, continuously tinkered for unexpected circumstances, and curiously written off as rigid and fixed—meshes with long histories of agile and adaptive statecraft in modern China. Political scientists Sebastian Heilmann and Elizabeth Perry argue that a principle of adaptive governance has been a durable feature of communist statecraft since before the founding of the People's Republic. Instead of locating political sensibilities as ironclad fidelity to one ideological doctrine or another, they approach political governance as a work style. They center a constitutional pragmatism to statecraft, traced to habits inculcated into revolutionary leaders during protracted periods of guerrilla warfare. "China's governance techniques are marked by a signature Maoist stamp" that approaches rule decidedly *not* as a caricatured authoritarian inflexibility, but as nimble adaptation to unexpected challenges. It is a "process of ceaseless change, tension management, continual experimentation, and *ad hoc* adjustment" (2011, 3).

Thinking of Chinese statecraft and rule as an experimental system is to center its capacity to change and be changed, and to understand the format of this change as a relation with the unexpected. The rules of the program are constantly being reconfigured as new challenges reveal the limits of any current version of the political. The state can thus be approached not through its overpowering ability to create worlds in its shape, but by centering its underdetermination and its flexibility when confronted with challenges.[40] The identification of late socialism with experiment forces us to train our attention on what Lisa Rofel identifies as the "tentative and somewhat ad hoc" nature of governance in contemporary China (2007, 7).[41] Drawing on Rofel, I pose the encounter of political and meteorological dynamics as a site where an encounter with environment—as both an object of power and an excess over it—throws this experimental impulse into overdrive. "Going green" is not simply an ideological program in the extension of an authoritarian mode of rule; it is an experimental state praxis in which "politics" is a challenge of becoming-with a changing nature. Environment configures a politics without guarantees, rather than a "green cloak to obscure [China's] current trajectory toward totalizing social control" (Li and Shapiro 2020, 34).

Governance is thus a disjunctive and ongoing process of experimentation, one that does not simply play out preordained internal political logics; it is shaped in complex encounters that exceed the state *per se*. The multiple temporalities, shaped in the embroilment of a mutable state formation and the dynamisms of *wind-sand*, offer a late socialist counterpoint to the clearly demarcated eras and phases of Chinese Communist Party historiography.[42] As Michael Hathaway writes in his discussion of the "environmental winds" that were sweeping across Southwest China in the early 2000s, new environmental formations cannot be smoothly traced from state logics and formations on their own. "The metaphor of winds," he argues, "suggests that we cannot know what happens by only studying Beijing's political proclamations." These proclamations must be situated in a social, environmental, and material-semiotic field that includes "all those who have a stake in what happens next" (2013, 11).

Late-socialist experimentation turns the question of the political, in the abstract, into questions of strategy, redeployment, and reconfiguration—specific ways of negotiating life in strange weather. In this light, state engagements and mutations with *wind-sand* are not simply an extension of a "war against nature" (Shapiro 2001) by other means.[43] Nonhumans are already ubiquitous in the shifting landscapes of environmental politics, and not only as passive objects to be eventually conquered. The dynamism of nonhumans is a point so fundamental that, for the environmental engineers, herders, city dwellers, and scientists who populate this book's airstream, it goes without mentioning. State scientists and environmental engineers charged with reengineering the aerodynamic topographies of dunescapes *cum* dust storm source areas readily understand that their everyday practice is not a problem of imposing an unchanging political will on a recalcitrant environment.

Instead of posing the political in the agonistic terms of struggle, grounded in the fixed poles and ready-made slots of an anachronistic nature-culture binary, I explore the interface of the political and the meteorological through experimental entanglements. These experiments do not lament over the failure of political formations to capture the wind. They are the sites where the actual work of political emergence happens, occasioned precisely by recognition of the limitations of extant political formations in the face of dusts and winds. I am less interested in asking how weather has *become* political, as if the contribution of critical work could only be in showing that some domain once thought to be neutral is *actually* political. In *wind-sand* as a figure of process, substance, and relation—one that entrains and displaces the political geometry of subjects and objects, the power that acts and the nature that is acted upon (see Strathern 1980)—we glimpse the possibility of another accounting for power.

Parallax Views

This book's insistence on staying close to *wind-sand* as a guide necessitates ethnography itself as an experimental practice in relation and perspective. What I call ethnography here is a series of ways in which the presumedly self-contained ethnographic subject is exploded then suspended into the weather. Ethnography, here, is an exhortation not to hold tempests in teacups, but rather to be swept into them, and to struggle to understand the self-enclosed domains of culture as part and parcel of worlds and planets still in the making. Ethnography thus is a way of relating to would-be informants not in the scene of intersubjective encounter focused on dialoguing human subjects, but rather, as cohabitants whooshed unevenly into a shared meteorological formation.[44] Following Andrew Mathews, ethnographic work begins in hiccups: I "hesitate in disconcertment as I [encounter] beings that I can only partially describe" (Mathews 2018, 391). This hesitation becomes a willingness to experimentation, where one's approach is inflected by its inability to cohere into a single perspective or explanatory framework.

One word for this displacement of anthropological knowledge practice by those of one's interlocutors is *parallax*. Parallax is the effect of optical depth when the "same" thing is perceived from multiple, nonoverlapping perspectives. Faye Ginsburg, in her reflection on the cognate and disparate genres of ethnographic film and aboriginal media in Australia, argues that in the juxtaposition of practices that both invoke and depart from one another, "one can create a kind of parallax effect." This effect can be harnessed analytically to "offer a fuller comprehension of the complexity of the social phenomena we call culture" (1995, 65). On reflection, it is my contention that the cultivation of such parallax, which, for Ginsburg is an ethnographic surprise, might be posited as an apt description of the ethnographic method as such.

Parallax, as a figure of offset perspective, conjures a field of vision that is at once shared and estranged. This effect does not require radical alterity, but locates seeing this dynamic of shared disparation of perspectives as ordinary. "The eye[s] of any ordinary primate like us" (Haraway 1988, 581, *pluralization added*) see in three dimensions because they overlap and yet do *not* completely share the same field of vision.[45] It is the optical effect in which multiple views on the "same" object from offset perspectives disclose a "reality that is exposed through difference" (Karatani 2005, 3). It is the effect of depth that is only possible through the disorientation of vantages on a shared object: not a question of radical alterities, but of the irreducibility of near-misaligned perspectives.

An ethnographic form that works through parallax deploys relational *irresolution* as a mode of knowledge. Consider the satellite imagery that opens this introduction, and how it tantalizes as experimental landscape art. Gestated in technomilitaristic fantasies of unhindered, un-located seeing, it might be understood as a technological refusal of multiple vision. I follow Donna Haraway to locate it, to situate it as a perspective from an offworld somewhere. Recuperating its one-off universality into an intertextual history of planetary vistas, we can attend to its generativity, not only in enforcing a singular knowledge, but in its potential uptake as an opportunity to open other senses of the planetary.[46] Even in NASA's description of the dust event, satellite imagery is both supplemented and estranged by eyewitness descriptions of the storm as an anti-optical event. The "god trick" (Haraway 1988, 581), its premise of omniscient vision, hiccups in this juxtaposition. The persistent anthropocentrism and ableism of the optical paradigm is perturbed in juxtaposition of human and satellite visions, as they do not quite coalesce in a way of attending the storms that "builds on our embodied capacity, but deploys it in a space that is utterly out of reach for any human body" (Ballestero 2019c, 780). The vertiginous parallax between multiple vantages both foregrounds and eclipses sense as a human capacity alone: the occluded optics and contracted visibility of the storm's morning-midnight and the topographical map of its meteorological landform from thousands of kilometers above.

The regime of perceptibility (Murphy 2006) specific to parallax is an emergent property of how myopias interleave with one another—of the generous discord that only disparate perspectives can make together.[47] Sensing through parallax is insisting that knowledge, anthropological and otherwise, happens *because* of multiplicity and not despite it.[48] To think through parallax as an entailment of ethnographic practice means that we are not gathering perspectives that will cohere into a single stable truth, but rather, generating a depth that is only possible through the juxtaposition of accounts and practices and worlds. This is also a kind of reflexivity that operates through relational displacement. It helps me to imagine ethnography as not just the enterprise of meticulously describing a coherent (human) lifeworld, but as the cultivation of dimension through the juxtaposition of multiple perspectives on the "same" (more-than-human) thing.

My own sense of what *wind-sand* is and how it matters was constantly being jostled in interviews and fieldwork with scientists, engineers, and herders, not because of their radically different understandings of what *wind-sand* was, but because of how close they were. In doing fieldwork and writing afterward, I came to understand ethnography as an exercise in gathering and juxtaposing subtly different perspectives on the "same" object, and in the stereoscopic multiplication of perspectives, generating a depth of understanding through the irreducibility of near misses to one another. I came to understand ethnography as a work of precipitating this parallax movement. Interlocutors do not only provide information or the requisite ethnographic sense of having "been there." They also participate in the continuous destabilization and reconfiguration of knowing, part of an experimental system that takes shape through the *mis*match that shared interests make evident. Engagement with the natural and applied sciences, arts, and humanities is critical and necessary. And interdisciplinarity is implied, not to build a cumulative argument, but to stake out positions whose differences will propel the movement of thinking.

We are after an optics of minor estrangements. In the destabilizing near miss of multiple viewpoints, any and every world appears as an opening, an arrangement at the edge of becoming something more than itself. Each arrangement of elements is populated with the possibility of others. Marilyn Strathern writes of anthropology as a cascade of forms that might propagate across domains: "Anthropological knowledge offers a transparent example of the process involved in rethinking through concepts and images whose expressible forms already belong to other repertoires and thus to other specific domains of ideas" (1992). Or, closer to the materials at hand, the movement of knowledge is powered by imbrication in densely material circumstances and finding one's senses shaken into other shapes in the encounter. Transformations in turn transform: as Mao Zedong writes, "If you want to know the taste of a pear, you must change the pear by eating it yourself" (2017).

AEROSOL FORMALISMS

The political meteorologies that concern this book enjoin me, again and again, to approach the very shape of the book as an aerosol form, taking *wind-sand* as both its matter of concern (Latour 2005) and the formal substrate of its literary morphology. I derive the notion of an aerosol

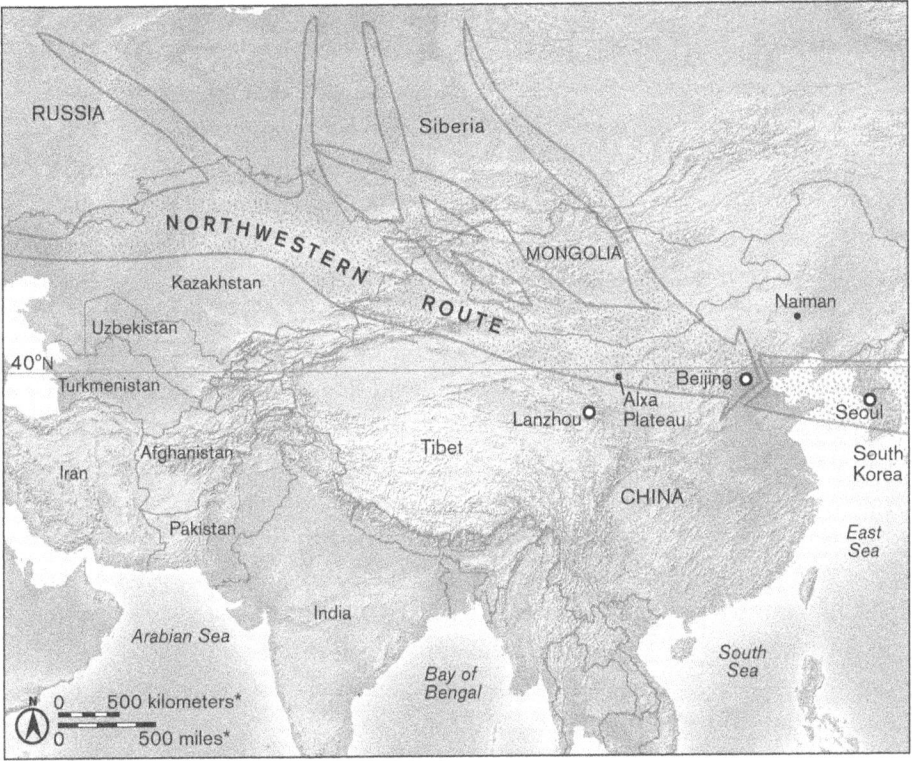

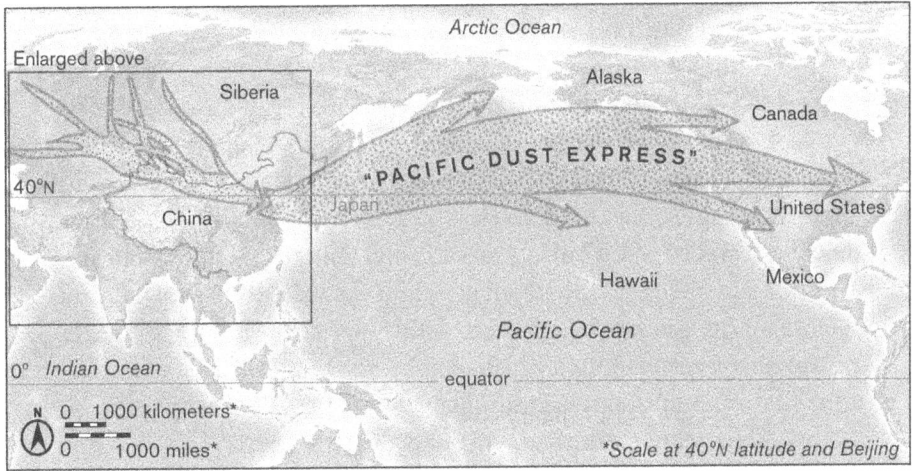

Map 1. The northeastern route of sand and dust transport, through and past Beijing.
Cartography: Ben Pease.

form from Theodor Schwenk's magical description of the flowing formations that emerge in the dynamisms of water and air. An "aerial form,"
argues Schwenk, is an emergent gathering of elements into the air,
whose dynamics and shape cannot be discerned except in relation to the
air that tissues them (1965, 116).[49] Schwenk's poetics of air allows us to
attend to the air and all that composes it as a process of formal emergence and change.

If the commonsense notion of dust poses it as chaotic, un-composed
matter—a dissolution of form or actively anti-formal—here I insist that
wind-sand exists through a profusion of forms. They orient a concrete way
of thinking about the planet without defaulting to the ubiquitous scale
of much of planetary thinking.[50] This is a dangerously attractive kind of
scale-making project (Tsing 2000) that we would be right to notice and
remiss to repeat. The multiple vantages on the mobile locality of aerosol landmass subvert the easy universalisms through which the planet
becomes a placeless everywhere, on the one hand, and the gravity of an
ethnographic tradition that desires to fix things in place, even as the world
and planet are on the move. Forms propagate (Strathern 1988): dusts,
politics, ethnographies, and planets are aerosol forms, concrete formats
of relation with and through *wind-sand*.

Desert winds create experimental worlds. They make airstreams into
architectures of relation and vectors of geo-meteorological assault. A map
of Beijing's 2001 dust events, produced by the Chinese Academy of Social
Sciences (fig. 3) through back-trajectory analysis, traces storm paths five
days back from skyfall at city limits. It reveals the city and its vulnerable
airspace as a skein of multiple "sand and dust transport routes," color-
coded into conventionalized paths. At the overlapping of colored airstreams and yellow sand sources, deserts and land degradation appear as
earthly precursors, forcing both analytic and state attention on a political-
meteorological process that connects lands and airs over long distances.
Weather maps encode the turbulent confluence of political concern, environmental onto-epistemologies, and meteorological trajectories into a
concrete itinerary of political and planetary emergence.

Such mappings invoke existing cartographic forms, like the self-
contained and self-evident territorial nation-state, while also opening
their presumptions of closed, flat space. Weather systems refuse national

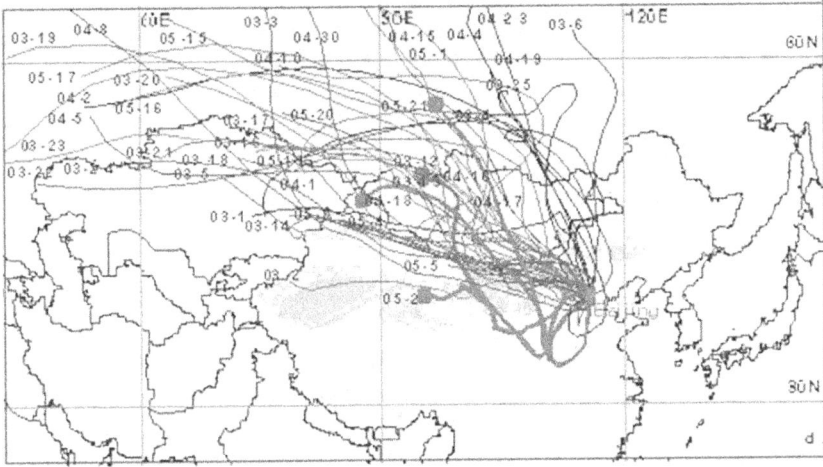

图 6　2001 年春季，北京所有高粉尘日 5 天前的空气气团轨迹。黑色和兰线代表沙尘输送的 "东北部沙地路径" 和 "北部蒙古路径"；"北部沙漠路径" 和 "西部沙漠路径" 分别由褐色和绿色线表示；粗红线代表的是 "转向路径

Figure 3. A Chinese Academy of Sciences map of the Beijing Spring, 2001. Each line traces an airstream five days before it reaches Beijing. From Chinese Academy of Sciences, "New Progress on Important Question in Asia Dust Storm Research," April 10, 2003, http://www.cas.cn/xw/zyxw/yw/200906/t20090608_601960.shtml.

borders. The rut of dust flows "makes and remakes its channels" (Tsing 2000, 327), substantiating seasonal dust-transporting airstreams as itineraries of material, meteorological, and political relation, concrete formations of an idiosyncratic planetary situation. The main period of research for this work was fourteen months of ethnographic fieldwork at various points along the dust stream designated on the weather map as an official Northwestern Route of sand and dust transport. It moved between Beijing and its dust storm source areas in desertifying regions in Inner Mongolia and Gansu as they were incrementally refashioned, through experimental social and landscape interventions, into a key zone in a provisionally reengineered weather system, centered on the capital. Moving then past the map's limits at Beijing's airspace, I traced the Northwest Route as it became a regional and eventually hemispheric avenue of atmo-terrestrial and political relation.

As my fieldwork proceeded, I had to learn to forego the identification of a country with its landed geobody (Thongchai 1997), so, instead, the book compiles an aerosol China, across phases and disparate parallax attunements to *wind-sand*. They required me to see China precisely as it is displaced and disoriented, far beyond the cartographic fiction of stable country-field that corresponds with others in a jigsaw national geographic (Malkki 1992). With *wind-sand*, China is a sequence of up- and downwind elsewheres, strung together through currents of aerosolized matter. Dustflows thus offer a wind-map for multi-sited ethnography. They generate "paths, threads, conjunctions, or juxtapositions of location in which the ethnographer establishes some form of literal, physical presence, with an explicit, posited logic of association or connection among sites" (Marcus 1995, 105).

I insert this caveat: while the book's field sites are multiple, its field is singular: a Chinese weather system, sprawled over spaces, jurisdictions, histories, and states of matter. In addition to fieldwork and interviews with state forestry officials, environmental engineers, and environmental scientists in China, I also spent long periods of time in tree-planting pilot schemes and construction and demonstration zones for sand-holding engineering works, resettlement villages for ecological refugees, and anti-desertification ecological research stations in Inner Mongolia Autonomous Region and Gansu Province. I worked as a volunteer for environmental NGOs in Beijing and Seoul; and I conducted fieldwork with joint teams of Chinese and Korean tree-planters along a regional dust-stream. I was fortunate to do follow-up research in the summers of 2016 and 2018 in China, as well as new research in California between 2015 and 2019.

Like *wind-sand*, the ethnographic form is subject to phase shift. It proceeds precisely by disrupting itself. The sections of the book condense around phases and stations in *wind-sand*. Between chapters, short passages I offer as *Apparatuses* puncture gathering narrative flows, torquing the shape of the analysis. When arguments seem to be settling in a fixed pattern, these apparatuses twist them into another. Across phases and hemispheres, the book turns with the kaleidoscopic permutations of *wind-sand*.[51]

Part 1, "Wind-Sand" explores the encounter of wind, sand, and state in China's continental interior. In a geography of frontiers that runs roughly parallel to the Great Wall of China, the antidesertification work deemed necessary to "protect" developed coastal cities from dust storms has largely occurred under the rubric of the Three Norths Project, a Great Green Wall of afforestation megaprojects that is to span the North, Northeast, and Northwest of China with billions of trees (Hong 2015).[52] As wind recuperates sandy lands into the dust-shed of the Chinese capital, they are increasingly seen in terms of their potential for motion by local bureaucrats, state forestry officials and engineers, environmental scientists, and populations of herders.

Sand stabilization, antidesertification, and other forms of environmental engineering reconfigure late socialist political institutions into experimental apparatuses for intervening in the geophysics of the Earth's surface, where wind and sand touch. As air and earth interact, vast tracts of semiarid China are becoming experimental zones whose ecological construction is an emergency measure to protect Beijing. The geophysical dynamics of *wind-sand* offer a profoundly materialist format for exploring the ongoing experimental reconfiguration of an array of late socialist governmental institutions.

Chapter 1, "Machine Sky," explores how state programs for dealing with inland land degradation are being reconfigured as programs of prospective meteorological engineering for the capital downwind. In the technical and governmental reframing of upwind land degradation as a matter of downwind dust storms, sky and land are appearing, both physically and in governmental calculation, as a continuum of substances in a single, complex process of meteorological transition. I explore the transformation of the social and terrestrial landscapes of Inner Mongolia as sociotechnical practices for engineering a machine sky, in which upwind places and people are figured as part of the geophysical machinery of downwind weather. As local officials—two days ahead of Beijing by wind—tinker with dunescapes and ex-herders, they offer a view of governance both as a way of securing the relationship between centers and peripheries of state power and as a continuous modulation of the relationship between sand and wind. This tinkering work is a meteorological reimagination of the very conditions of statecraft, enacting late socialism and its institutions, earthforms, and populations as a machinery for transforming the sky.

Chapter 2, "Groundwork," builds on the previous chapter to explore late socialist experiments in ecological governance as they take shape around the holoparasitic botany of two desert plants: a windbreak and its potentially valuable rhizome parasite. As state forestry institutions shift from the provision of timber resources to the planting of vast tracts of windbreaks, they reformat state institutions of socialist redistribution, economic development, and social control into a complex bureaucratic machinery for shaping the economic and affective environments that will compel herders to turn to forestry. The botanical relation of two plant species thus emerges as a general template to coordinate ecological and economic change, rearranging late socialist rule through the relations of plants as a mode of environmental infrastructure (see Hetherington 2019).

If the first two chapters elaborate, respectively, a reconfiguration of governance through *wind-sand*, and then a specific state project of landscape stabilization, chapter 3, "Holding Patterns," witnesses how the horizons of sand stabilization and dust storm mitigation are undermined and reshaped with the sandflows they seek to contain. It explores how state sand engineers and ecologists in the projected trajectory of advancing deserts learn to tell time through sand. If the environmental present, under the sign of the Anthropocene, has been figured in a convergence of human and planetary timelines, sand girds provisional modes of geological and political temporality. Sand materializes here not simply as a substance to be controlled, but a chrono-political medium through which multiple timelines take shape and contend. The manipulation of sand appears thus as a manipulation of time, plotting out minor genres of environmental time that I call holding patterns, marking time against the inevitability of future erosions.

Part 2: "Fine Particulate Matter," follows *wind-sand* from continental cradles of dust storms to the center of China's atmogeography of meteorological insecurity. This section considers the physical, technopolitical, and cultural enactment of Beijing's air as massively dispersed semi-solid. Dust and particulate matter, I show, reveal Beijing through disparate struggles to weather the density of its atmosphere. Quotidian artistic and political experiments, entrained into the fluctuating concentrations of fine particulate matter in the atmosphere, make urban life evident through the reworking of bodies, airspaces, and polities as peculiarly meteorological in character.

Chapter 4, "Particulate Exposures" zeroes in on catastrophic events of particulate density in Beijing, exploring how figures of breathing as a mode of exposure have proliferated in the city. The incorporation of breathing bodies into the aerosol dynamics of the city's air offers a space to consider a becoming-atmospheric of the collectivities, contestations, and horizons of late socialist politics in the city. We work through scenes of exposure that coalesce urban life in conditions of quotidian necrosis, without either the horizons of insurgent life or resigned death. That breathers are "continually forced to breathe" (Zimmer 2020) reveals bodies into political possibility in accounts of happening that do not fall easily alongside the accounts of agency that are available either in the liberal notions of civil society that some associate with air pollution politics or in phantasmatic revivals of a Maoist People.

> those who live
>   are continually
>   forced to breathe[53]

While state media attempts to stabilize embattled breathing into injunctions to sympathize with state power, other meteorological and respiratory figurations of collectivity take shape. Socialities and relations with the state are recompiled with the biomechanics of urban air. How does the demand for respiratory empathy become an injunction to identify with power, against the oblique possibilities of breathing collectives in late socialism?

   Chapter 5, "City of Chambers," explores technical and quotidian interventions into particulate matter physics to elaborate a poetics of urban space. It focuses on the technologies and techniques of airspacing that render urban life a play of voluminous dimensions and gradients of dust. As a particulate rejoinder to Michel de Certeau's (2013) meditation on walking in the city, particulate predicaments unfold an urbanism of "pedestrians of two flat dimensions" (Serres 2012, 136) into a city of chambers and volumes. Airspacing, the realization of a city into a morphology of adjacent and disconnected airspaces, is an aerosol technics and poetics. Its desire for respite in the air from the air can be traced through attempts to hermetically seal atmospheres against one another—a desire continually frustrated by aerosol infiltration and the persistent porosity of bodies and spaces. In its second half, the chapter explores two art projects, both of which address the air as a solid that can be captured and precipitated into particle solids, which become fungible with square kilometers of civic airspace or weeks of particulate matter.

Part 3, "Continent in Dust," moves back and forth across the Chinese border and then chases dust downwind as it blows across the Pacific. It explores how "contemporary cultures of anxiety" (Byrnes 2020, 2) over China's rise are filtered, sometimes quite literally, through specific geopolitical and technoscientific practices that attempt to make sense of floating landmass and that stumble over the available languages for accounting international relations along modern weather systems. Incoming Chinese dusts thus foment an array of technical, scientific, and ethico-political practices that wonder and worry over the rise of China as a matter of both global and planetary concern.

Chapter 6, "Downwind," returns to questions posed in this introduction by inquiring, ethnographically, into the parallax between geopolitics and geophysics as dusts cross international borders on the hemispheric airstreams that NASA has called the Pacific Dust Express (Barry 2001). At each station of the weather, it considers downwindness as it gathers and deranges significance, shifting from a seasonal geographic fact into an array of ethical, affective, and legal disorientations. South Korean tree planters organize planting trips to dust storm source areas that are upwind of both Seoul and Beijing, and negotiate interdependency, vulnerability, and culpability in times of bad weather by insisting on a kind of friendship through dust. US-based scientists precipitate long-traveled Chinese geology out of air at the Pacific edge of the North American continent to argue that particles of Chinese land are now the baseline for American air. In particulate traces of aerosols impacted on Mylar they search for a continent in samples of incoming dust. Between these multiple accounts, we explore China's rise from downwind as a contention of interwoven weather- and world-systems.

MACHINES FOR A CHINESE PLANET

A collection of Apparatuses sputters between the main chapters of this book. These apparatuses are an internal paratext running both parallel to and askance of its main arcs. They serve as interludes, and, together, they make up a little chapbook. Their purpose is to disrupt the book's argument as it seems to coalesce, and thus approximate a literary form for parallax thinking. They bridge chapters by estranging them from one another, mimicking the jostle of a kaleidoscopic pattern into other arrangements. Some of these apparatuses are machines; others are political philosophies and clichés turned speculative anthropology (Whitington 2013). Poems, composed or accidental, recur. They are all Apparatuses, and I offer them as exhibits in a musée mécanique, or beasts in a garden of Anthropocene—or Sinocene—devices.

Experimental apparatuses stage worlds only to reconfigure them. In this sense, they are anthropological devices *par excellence*. They demand a defamiliarization of existing worlds and a sensitivity to the possibility of other ones. In the manipulation of matter and concept that they make possible, they are onto-epistemological devices. I draw this understanding from Karen Barad, who defines apparatuses as "complex agential intra-actions of multiple material-discursive practices" (2007, 140). In apparatuses, knowledge emerges through practices, "specific material engagements that participate in (re)configuring the world" (2007, 91). In not just acting on the world, but participating in its reconfiguration, apparatuses—such as the wind tunnel and its cohort of others that bridge chapters in this book—redirect, supplement, and teach new modes of attention.

Apparatuses concretely stage the limits of ethnographic knowing, and thus are taken up as horizons for the reorganization of sensory habits. Tim Choy describes research assistants who must train themselves to register the subtlest distinctions of elusive mushroom scent, which are chemically separated into volatile compounds by an experimental device (Choy 2018). Their capacities for noticing are still in process, oriented toward the still indiscernible gradient of scent and sense precipitated by the experimental apparatus. The apparatus thus offers the researchers a glimpse of what they are not yet, but may become, able to notice. In training their capacities into harmony with the demands of the mushroom's experimental apparatus, the researchers thus find themselves retrofitted by their machineries as they seek to extend their sensibilities across this newly realized gap.[54] The apparatus is one anchor of a generative epistemic and practical relationship, out of which a given understanding of the world is shaken, and patterns of thinking diffracted through its contingent ensemble of moving political, technical, and material parts.

Out of a staid now, these apparatuses foment multiplicity and possibility, coaxing phase shift out of a solid world. The "ontological incommensurability between human and machine" (Fisch 2018, 8) in these encounters stages a way of sensing beyond oneself, of thinking toward what cannot yet be apprehended. Or, better, they offer a machine diagram for recircuiting one's sensory and analytic habits. Apparatuses then allow for a collaborative displacement of sense and conceptual habit. The technopoetic, cyborg attunements that they invite extend and undermine ethnographic authority. This is a phase shift, too. Like *wind-sand*, with *wind-sand*, we are decomposed ever into another composition.

PART I   Wind-Sand

tirelessly building
as wind builds, rising
as wind rises, steadily gathering
to nothing, quiet, and
the wind rising again.

Li-Young Lee, "Furious Versions"

In the wind tunnel of the Chinese Academy of Sciences' desert research labs in Lanzhou, aeolian physicists prepare to build a dust storm.[1] They load samples of sandy soil gathered during summer field research into one end of the long, Plexiglas tunnel.[2] With a special rake, they shape this pool of sand into a flat sheet, the bed of a tiny, purpose-built desert. Out of sand carefully collected and processed *in situ* then shipped to Lanzhou in burlap sacks, they build a metonymic enactment of the state-designated dust source area (*fengsha yuandi*) that the sample comes from. Activated by computer from an adjacent room, the wind tunnel's large wooden fan begins to turn. It gains speed, pouring air in a flowing stream over the carefully prepared dirt.[3]

In the tunnel, wind and sand drag into the substance that the physicists and their colleagues at the environmental engineering research institute that houses the tunnel call *wind-sand*. At first, the fan-driven wind passes over the sandy bed, a sheer force skimming and scraping its surface. This is an in-between time, where the earth and air of this experimental desert remain a terrestrial and an atmospheric volume, adjacent and separate at the sand's surface. As the fan-blown air continues to work over this recreated desert, however, the sand begins to quiver, as particles entrain in the air's traction. In motion, the air and ground shift from discrete spaces into interacting substances. The miniature horizon that divides them becomes a site of geo-atmospheric activity: "aeolian processes."[4] The flat sheet of sand looses into motion, the topology of its wind-exposed surface catching and warping with the push of the air. Grains jump in the wind before falling back to the simulated slice of China's landmass in the tunnel. *Saltation*: a leaping of grains. Sand particles kick up from the sand bed and bounce back into it, transferring their propulsive force to kick up even more at the craters of their impacts. Then, *suspense:* dusts lifted and shot into the wind stream into fine particulate emissions, an aerosol suspension that holds until the swift grip of wind loosens land back to ground again.

The tunnel's desert lurches and heaves. It bangs and flows down the length of the tunnel's clear shaft as a contained weather system, which is in turn translated into cascades of data and photographic stills: basic science for a fluid earth. It crashes against the tunnel's clear plastic walls and across the thresholds that cleave apart states of matter. The desert's protean topography, between surface and air, is a collaboration of *wind-sand* and the technoscientific enactment of state concern, shaping on the tunnel's bed into alluvial ripples and eddies in the river of fan-released wind. Dusts rise, a simulated landscape methodically blown into a simulated storm. Between saltation and suspension, the impacts of leaping grains blur the surface into sprays of experimental materiel, carefully photographed and recorded for further analysis.

The central wind tunnel in Lanzhou and its associated constellation of field stations across sandy China anchor a distinct historical ontology of the country.[5] The wind tunnel runs through the winter, making dunes and dust storms out of piles of sand collected by science teams during summer field research. The tunnel enacts its "China" as a gradient of terrestrial susceptibilities to flight. If the collected samples of dirt metonymize their respective sandy lands across the country's semiarid north, the tunnel unfolds these landscapes into a typology of *wind-sand* potentials. In *wind-sand*, two nouns abut. Earth and sky contort, with the collision of wind and sand, into a generative grammar of substances and transitions. In the tunnel and the vast scientific, political, and technical assemblage structured through it, simulated Chinese land and air give way to a vision of a country unlatched from the firmness of its ground. Its China is made and unmade in sprays of *wind-sand*: a substance that is a relation that is a process that is an event.

Out of their preparation of field samples of national dirt, the aeolian physicists compose the country as a mosaic of aerodynamic earths. One storm at a time, the physicists stage another China: one in which the terrestrial guarantees and presumptions of territorial sovereignty roil into a theater of substances, earthly fluids, and phase shifts. The tunnel presages a cartography of the country as a riot of *wind-sand* conditions. Place is not a *here-now* but a *where-through* of aerodynamic land and hypothetical winds. In maps plastered around the institute, places are sprays of arrows that map potential stormpaths. Mountain chains are recontextualized as dust channels and the broad plateaus of northern China as the dry beds of rivers of dust. In the tunnel, the *terra* that girds territory, terrestriality, terrain, and most importantly *terra firma* disjoins the earth from the comforting expectation of solidity.[6]

The wind tunnel is a proposition for a way of thinking of land and air in which the divide that separates them opens into an interface where one permutation of *wind-sand* gives way to another. Not only do sand and wind physically reconfigure in phases of dust, of *wind-sand*. So, too, might a metaphysics of the earth that prioritizes fixity and not movement, solidity and not the thresholds, transformations, and unfurling of planetary and political matter.

At the "encounter of air and apparatus" (Choy 2018, 55) the tunnel setup bursts the isomorphism of land and landed-ness, revealing land as process and event to be activated on a felicitous wind. Its fan sets into motion a dance of land and sky as interacting substances. Terrestriality becomes a capacity to meteorological permutation, a becoming-weather whose physics will make worlds turn. The experimental apparatus orients toward land as extraterrestrial, a tempestuous potential toward becoming more-than-land. It is registered, in the tunnel and in the China generated in the tunnels' walls, as a tendency to phase shift. Between sand and *wind-sand* are storms lying in wait for a propitious wind.

1 Machine Sky

SOCIAL AND TERRESTRIAL ENGINEERING
IN A MODERN WEATHER SYSTEM

The riverbed in which power runs leads back, the patterns
of power lead back, the valley where power is contained
leads back—to the forever new, endless, straightforward
way. Reversal, recurrence, are the movement, and yet the
movement is onward.

Ursula K. Le Guin

Rolling dunes undulate wildly outside the windows of the shiny Chinese-made jeep that Forestry Secretary Xiao has claimed for the week. From the smooth new government road, the landscape passes as a frenetic rhythm through the tinted side window. The tar-black of fresh-laid asphalt jars starkly against the muted colors of the desert springtime, hazy yellow in the transition from winter to spring, the year's most intense season for dust weather. Craning his neck toward me in the back seat, Xiao explains, "This area, because of its proximity to the Helan Mountains," an inland range crested by the serpentine western reaches of the Great Wall, "receives much more water than deeper in (*litou*)." By *deeper in*, he means the yellow streak of dunes extending toward and past the horizon: the vast belly of China's continent, as though the jeep were skating on the rim of the Earth. "That is why we can seed here."

We are on the second of a days-long trip across the Alxa Plateau in western Inner Mongolia, surveying a series of dust-control engineering projects that will lead us from the region's main city deep into the sands and degraded pastures that surround it. Even on fresh road, the driver abruptly veers, sometimes driving for stretches on the gravel shoulder to swerve around sand drifts that have blown onto the new roadbed. Sand creep makes the straight new road an erratic, curving course. So the long drive is studded with long breaks. We step out of the jeep to catch our bearings, into the lazy peals of dust that rise and break off the road's edge. Xiao ceremoniously offers cigarettes to me and the driver, then struggles to light his own, shielding the lighter against the stiff, constant wind. We wade into an expanse of shrubs at road's edge, taking care to avoid smashing fresh, tender shoots.

Looking away from the mountains we see shrubs and sand make a chromatic gradient that fades toward the *inside*. Bursts of scrubby shrubs radiate out from the edge of the fresh asphalt, slightly tacky in the mid-day sun. Their seeds have been dropped over multiple years from the holds of crop-dusting airplanes that the forestry bureau hires for aerial seeding campaigns. Early spring buds dim into silver-green fronds. Their colors pixelate into the yellow of the open sand farther and farther from the road's edge. The gaps between high dunes funnel wind into jets that swish the earth into sky in twisters that rise and break. Green, gray, yellow. Shrubs, sand, haze.

Leaning against the jeep's hood, the driver geolocates us with a handheld GPS device, studiously jotting down coordinates in a notebook he keeps in his shirt pocket. While he does this, Xiao details the forestry bureau's work, pointing with the cigarette across the landscape as though describing a massive environmental machinery. The forestry bureau, he explains, hopes to reconfigure the geophysics of mobile desert at, above, and just below the sands' shifting surfaces. In its work of planting, building, and seeding, the bureau is in the process of reengineering the landscape to intervene in the flow and movement of wind and sand.

The shrubscape doubles the road to mark a doubled frontier, an ambivalent relation to place (fig. 4). The road is a line in the sand. For Xiao, it is also a line in the sky. At its edge, we face the airdropped architecture of windbreaks and sand swales that the forestry bureau hopes will divert, pool, and pile blowing sand. Facing deeper in, our backs are toward the Helan Mountains. And just beyond that, as the driver casually mentions, the open east of the spring wind stretching toward more important places than the sparsely populated plateau: Beijing, hundreds of kilometers and two days as the dust flies on a strong spring current. Should sand from these dunes whip high enough above the surface to catch into the wind

Figure 4. Blank sign with *suosuo* plantings and trampled wire. An ambivalent place: disappeared or yet to come. Photo by Micah Hilt.

and breach the mountains and their crumbling, ineffectual Wall, this land will arrive over the capital as a bout of bad weather.

Too much of this earth, Xiao jokes, floats into Beijing's sky. The land degradation that has exposed surface sands to the wind have, for the local forestry bureau under central direction, emplaced Alxa into a geography of dustflow traced back from the capital. Xiao's forestry bureau has a mandate to lead the transformation of sandy land against its insistent tendency to suspension on winds. The bureau's local work is part of a much broader ecosystem of programs for reverse engineering the particulate densities of Beijing's atmosphere. The bureau oversees the reconstruction of dunes, degraded pastures, and transitional desert to intervene in the complex geophysics that will eventually make this place into downwind weather. For days, we will pass through the plateau as its landscape unfurls in a vast construction zone of barriers, channels, and windbreaks, each of which is a small monument to the meteorological anxiety of the capital, a straight shot by wind from here.

Years of seeding have played a part in transforming open sand into a desert earthwork. The ongoing intervention of the forestry bureau must be understood "not as initial impositions of form but as mere redistributions of matter" (Gladstone 2019, 394) that, over time, will not stop the wind wholesale, but will change the way it interacts with the sand. Seeding and a battery of associated social and landscape interventions, they hope, will modulate the aerodynamics of the plateau's sand.

In concert with a network of local state institutions, the forestry bureau has also been central in leading experiments in the governance of rural life in Alxa. New investments—and also divestments—in the location, behavior, and livelihood of rural populations, approach social governance of herders and ex-herders in Alxa as a key part of reengineering the geophysics of local sand and Beijing-bound wind. The management of both social and environmental processes has become an experimental terrain for reengineering the weather.

"We cannot stop the wind," says Xiao. "We can change the way it moves across the land." The shrubland and its wind, assures Xiao's bureau, will change together. Between the interphases of sand and wind, the bureau seeks to create patchworks of stable sand and airflow, revealing both landscape and human life as engineering problems, part of a geotechnical assemblage that is "ecologically novel and constantly in flux" (Whitington 2019, 8). What shapes their interface at earth's surface and the routes dust might stream, the sandblasted riverbed in which power runs?

The infrastructures they install into the land, and their practice of building social infrastructures to facilitate the renovation of *wind-sand* dynamics do *not* seek "to remake landscapes and force the natural world to conform to these projects" (Larkin 2008, 36). Instead, Xiao's topographic intervention approaches landscapes as processual and in flux, disallowing the possibility of a single, final infrastructural fix. Where there is change, Xiao wonders how to properly enter the dance, situating the state, human sociality and motivation, and *wind-sand* together, in the thick of the storms that might yet be preempted before they spill across the North China Plain. He wonders how to change how things change. Instead of forcing the natural world to conform to state projects, he challenges us to notice how state projects are part of the chaotic social, geophysical, and political machineries through which weather systems evolve. Dunes and the spring wind against our faces; mountain chains, crumbling walls, and the centers of power at our back. The social and terrestrial landscapes of Alxa appear as the inner workings of a prospective national meteorology, traced backward from the capital through the storms that may yet come. It takes shape in one burst of brambles, one season of wind, one experimental resettlement village at a time.

———

This chapter explores the state-led reengineering of Alxa's terrestrial and social landscapes by detailing experiments in tuning earthly, aeolian, and social processes into a prospective storm protection system for the capital. It is organized as an inspection tour through the "ambiguously natural and crafted" (Haraway 1991, 149) engineered worlds of a cradle of dust storms.[1] Here state forestry agencies are busy in their attempt to reengineer its terrestrial and social worlds into the functional mechanism of a machine sky. In this, the chapter attends to forestry bureaucrats as they pose the environments and worlds of late socialist Inner Mongolia as moving parts in the geophysical machinery of continental weather.

Forestry officials in Alxa have experimentally reimagined social and environmental governance as a matter of recuperating social and geophysical dynamics into meteorological mechanisms imagined across national scales and relations. I approach this process of geophysical and aerodynamic landscape tinkering through the experimental modulation of what I call a "machine sky." With the notion of a machine sky, I am pointing to an emergent conception of meteorological dynamics as the complex and cumulative effect of entangled environmental, political, and social processes across vast distances. The operative and experimental dynamic at the center of the notion of a machine sky is that the contents and dynamics of key weather systems are being enacted as giant and complex geophysical machineries that require engineering, maintenance, and intervention. In dust storm mitigation, this has meant that governance in upwind places is being recalibrated into questions of geophysical emergence and meteorological relation.

The architecture and argument of this chapter replicate the way that the organization of local environmental engineering projects hold, in their distribution across the plateau, the shape and movement of dunes and dusts. Local and visiting bureaucrats experience this political and geo-physical entanglement in space most viscerally in routine tours through, as it were, the guts of the weather machine taking shape across the pla-teau. We follow, in turn, environmental engineers and forestry officials racing into the wind in their attempts at modifying the aerodynamics of degraded, dust-generating land. This chapter then considers pilot pro-grams in ecological resettlement of ex-pastoralists. In these experiments, rural social governance appears as a fine tuning of state support for and neglect of ex-herders evacuated into state-built resettlement villages, which in turn becomes an important means of securing labor and materi-als for afforestation techniques like mass planting or aerial seeding. Mov-ing between dune stabilization projects, anthropologists and bureaucrats get an expansive sense of the strategies and experiments deployed for reengineering earth, sky, and their turbulent interfaces.[2]

The deserts of the Alxa plateau have, since the 1950s, been involved in state programs of political and ecological transformation. In the exuberant optimism of early communism and Maoist voluntarism, these programs approached sand as an engineering problem to be resolved through labor and technical ingenuity in conditions of scarcity (Lu 2006). Forestry, land management, and massive public works projects posed sand as a material to be controlled, conquered, and made productive. Deserts could be con-verted into productive oases for miraculous gains in agriculture and ani-mal husbandry in a heavily Mongolian-nationality part of China. Among other major national-forestry programs, since the late 1970s, coinciding with the advent of the Reform Era, Alxa's dusty plateau has been slated as an afforestation zone under the Three-North Shelterbelt, a massive anti-dust and desert-holding tree-planting and environmental-engineering project across the country's north, northeast, and northwest.

This shelterbelt is one among several national forestry programs that explicitly aim to remake landscapes across northern China into an inte-grated dust protection system for the Jing-Jin-Ji region, the national capi-tal region around Beijing, Tianjin, and Hebei province. This geography of shelterbelts has figured centrally in plans to keep dust out of Beijing's air, which have ramped up since consecutive seasons of intense spring dust storms over Beijing in 2000 to 2001. In popular shorthand, the massive shelterbelt project is called the Great Green Wall of China. It has been touted by the Chinese government as the largest environmental engineering project in human history, covering a full 40 percent of Chinese territory and, promising by 2050 to triple forest cover with billions of trees across northern, northwestern, and northeastern China that will radically reshape the topography of China's land and the routes of its winds.

Since the early 2000s, western Inner Mongolia's Alxa Plateau, has landed in an archipelago of zones for experiments in dust storm control scattered across Beijing's spring upwinds. As upwind territories in Beijing's seasonal dust shed become a constellation of dusty choke points (Carse, Cons, & Middleton 2018), modernist state desires to pacify an unruly nature have increasingly given way to political experiments that instead seek to mold new landscape formations with different storm potentials. In the reimagination of land degradation as an upwind meteorological concern, forestry efforts in northern China are shifting from grand plans for conquering deserts into experiments in long-distance weather modification. Dustflow presents, in these projects, as an engineering dilemma of national significance.

Diagrammed into the workings of this machine sky is a sense of landscape as a technical, processual, and political morphology. The fluid aerodynamics of these landscapes are in continuous feedback with the social, historical, and governmental conditions of Alxa more broadly, when figured as parts of a longue-durée terraforming process whose meteorological effects play out over much longer distances.[3] Environmental engineers thus enact the weather as a zone of "ideological and material encounters with traces of China's past that opened up new imaginings for China's present and future" (Karl 2020, 4).

Forestry experiments explicitly pose the social and physical landscapes of some places as moments in the weather systems of others. Embedded in this is a sense of Beijing's atmosphere and actual or potential weather events therein as the cumulative effect of political and environmental dynamics of upwind elsewheres. The contents of any "local" airspace are thus sourced to social and geophysical dynamics far beyond given administrative units tied to fixed territories. In turn, environmental engineering projects in Alxa are reimagined as part of a cumulative tinkering whose success can only be registered through weather events far downwind.

The notion of a machine sky here allows us to attend to the "engineering diagrams" that guide these social and environmental interventions. Diagrammatic thinking emphasizes the conceptual and formal sensibilities of these interventions. It allows us to attend to Alxa's sands and winds as zones of governmental experimentation aimed at proactive integration of "a multiplicity of disjunctive elements and spin out new relations between them" (Jensen and Morita 2016, 615). I am interested in how, in practice, forestry bureaucrats and engineers increasingly attend to social management and environmental engineering as elements in a higher-order meteorological process. They approach matter, and for that matter sociality, as "pregnant with morphogenetic capabilities" (De Landa 1997, 34) out of which new topographies and meteorologies might be shaped.

Officials like Xiao seek to devise means of terraforming new weather patterns out of the dunescapes. They do not aim to impose some prefigured order on the chaos of the world.[4] Rather, their purpose is to firmly insert state institutions and hopes into the planet's perpetual change: to strategically intervene in its physics to cultivate morphogenetic capabilities that remain latent.[5] In Alxa, environmental engineers and local bureaucrats approach the social and geophysical conditions of late socialism in the region—extreme and rapid land degradation and a state-led collapse of the region's pastoral economy—as precursors to a dust event. They see in the social and ecological "rubble" of rapid desertification (Gordillo 2014) materials at hand that they might shape into the technonatural infrastructure for a still-unrealized national weather system.

Xiao's forestry bureau's approach to landscape and social formations dispenses with any static distinction between what is properly natural or cultural, thinking of *both* society and geophysics as manipulable elements in an open-ended meteorological dynamic. Their approach is reconfigurational. It works through the redeployment of existing governmental expertise, attention, and tactics. In this, planted and synthetic infrastructures for desert-landscape engineering and pilot programs for the management—or as we will see, neglect—of local (ex-)herding populations are to be understood not as fundamentally distinct registers of political attention. Rather, they are to be tracked ethnographically as they are brought together in state landscaping projects as moving parts in a broader geophysical system. This is a confluence of meteorological and political formations in which "unexpected historical conjunctions fall into new coordinations" (Tsing 2015, 4) to change the shape and flow of land, air.

WEATHER MACHINES

Scientists, environmental engineers, and bureaucrats across the land and air of a Chinese weather system gather around the question of how to build social and physical landscapes that will yield better, more secure weather for the capital. Reprogramming, through a diverse set of interventions, the processual qualities of landscapes is the crux of an increasingly generalized political logic that understands upwind regions as an incipient infrastructure passing into regional weather patterns.[6] "Long considered essentially static," meteorological conditions once figured as background are "now regarded not only as naturally variable, but also as subject to alteration by human activity" (Edwards 2002, 194). In attempts at modifying dustflow and dune motion through the rewiring of Alxa's terrain, "weather" is no longer a passive background. Even if, as Xiao laments, they cannot stop the wind, through environmental engineering and new kinds of social management, they can modulate dusts. Alterability, figured as a principle for political action in geosocial and geophysical systems, turns the radical open-endedness of environmental processes from a limit of power into to a condition of its reconfiguration.

Machine images have long troubled boundaries between human, animal, and technology (Haraway 1991). They allow a glimpse into a mode of entanglement that centers the functional (or dysfunctional) relation between moving parts, drawing disparate elements together without tendentious handwringing over the integrity of staid boundaries between the natural, the social, and the technological. As Natasha Myers writes, machine images enact worlds in "three-dimensional, temporally dynamic dimensions" (2015, 161). In environmental history, machine images are deployed instead to point to the entangled and emergent character of environmental systems. For instance, in Richard White's history of the Columbia River (2001), salmon, nuclear power, dams, and water together make the river a single, internally complex "organic machine" that thwarts the boundaries between technical and natural domains. Planetary weather and climate systems appear as complexes of meteorological dynamics and the political and technical infrastructures for knowing it (Edwards 2013). Made and remade, these environmental machines are not defined by optimal operation but ceaseless change. They gather apparently incommensurable processes, ontologies, and concerns in a wild tangle of partial connections (Strathern 2004).

And in this tangle, the entirety of Alxa's social, ecological, and political life has become a set of geo-atmospheric precursors, points in a geophysical calculus that place the state squarely in the midst of worlds in formation. In an important sense, the "late socialist social" at play (Collier 2011) is itself part of this tangle. If infrastructure, following Ashley Carse is not taken to refer to any "specific class of artifact, but to a process of relation-building and maintenance" (2014, 11) that draws together disparate human, natural, and technological elements into a larger more-than-human and more-than-technical arrangement, we might locate state institutions within infrastructural relations and not just as the master-planning agent that builds them out.

This has important consequences not only for locating state practices as interventions *into* environment as some exteriorized reality, but also for figuring political action as always occurring from within the milieu of an engineered world already "in operation." Building an infrastructure into sand and wind drives cadres and engineers to passionately attend to how these substances interact and how to better interact with them.[7] It is not to explode their choreography but to enter their dance and reshape its movements and rhythms. The political reorientation to the country's changed environments as an infrastructural problem profoundly embeds the state *within* the machine skies that it seeks to transform, rather than simply acting into them as an alien presence, exterior to and extricated from the dynamism of the world.[8]

SCULPTING IN WIND

Far from the Inner Mongolian roadside where Xiao's forestry bureau sows seed out of low-flying airplanes, Deputy Secretary Luo flips back and forth through a deck of PowerPoint slides on dune restabilization through strategic planting. His space heater cuts ineffectually against the Beijing winter seeping through the windows of the state environmental training office, where he works as a trainer and fixer for visiting international delegations of antidesertification officials, mostly from sub-Saharan Africa. In the empty weekend office, Luo talks me through a training while he flies through the slides on his desk computer, filled with photos that he and his colleagues have taken at planting sites in Beijing's airshed: some in the immediate fortified-forestry zone that completely rings the city, and some in places as far afield as Alxa. The air outside the window is a white haze, thick with the smell of smoke.

At the center of my conversation and training session with Luo is a repertoire of strategic planting techniques for dune stabilization that the central government's State Forestry Administration has, in the last decade, promoted as part of a tool kit of international best practices. "Thirty or forty years ago, we thought we could just plant trees everywhere, and they would grow on their own and the desert could become farmland." He chuckles a bit before continuing, "Now, we have to be more strategic, we have to make decisions about how, including choosing the right species and location." The engineering approach Luo teaches starts from a practice of attention, scaling down to concentrate on principles of dune morphology, then local economic and ecological conditions, a movement that belies both the generic character of his argument and the resolutely nationalized stakes of dune stabilization as it scales up to patterning weather. Past antidesertification programs, in his thinking, are framed less in terms of their ideological excess and more in their financial, technical, and ecological feasibility, which he waves off as the wistful but naive adolescence of the country's desert management expertise.

Environmental engineers like Luo approach winds, barriers, and plantings as storm-sculpting tools. His pedagogical attentions turn on techniques that advocate reshaping the dynamics of surface sands and winds so that the processes that drive dune formation can be reworked into techniques of dust-storm abatement. This interplay of dusty scales, in which the largest storms originate in the aeolian physics of the smallest, lightest grains of mineral dust, finds engineers fixated on the shifty topographies of windy dunes, a peculiar geophysical kind of atmospheric attunement. Chinese anti-dust-storm forestry engineers taught me not to think of wind as only a force of destruction to be stopped. Rather, they attend to wind as a crucial component in the physical mediation of fluid sandy land and weather systems at ever-expanding scales.

The very multiplicity and power of the wind have shaped Inner Mongolia's peculiar place in the technical projection of particulate futures across China, playing out in a knot of adjacent and sometimes-mutually-contradictory aeolian geometries. They pose that winds at the surface, for instance, are a potent storm-making threat. It is the environmental medium that migrates particulate pollution like dust or coal smoke. Where dusts settle, residents wait for fresh winds to flush their air, pushing particulate matter further downwind. But higher up above the wind-whipped surface, strong, smooth winds are crucial to the freshly planted wind turbines that seek to bank on Inner Mongolia's newly discovered wind energy potential.[9] These turbines will make coal-rich Inner Mongolia a crucial site for envisioning the technical conditions of the energy transition that will keep coal soot out of urban air.

While the wind is continually identified as a major environmental problem, in Luo's description, wind is a force that shapes land and weather. Wind is an element in a sculptural topography, defined by the ways that it shapes itself over and over in time. In his landscape-engineering strategy, dust-kicking winds become also the most efficient means of shaping a landscape to dampen their very force. And as winds reshape a sandy topography, the topography also repatterns airflow. In this sense, for Luo, landscape is not a fixed view, but a process of continual change in the interactions of wind and sand in a dance of morphologies. One must understand the technique he teaches as a means of recuperating the wind's tempestuous power to drive the engine of an engineered landscape machinery.

While from Beijing the dust problem appears as a series of smooth aeolian vectors that link Beijing to the dust source areas that radiate upwind into the Asian continent, Luo's technique begins from *wind-sand* microphysics to chart out a dense profusion of aeolian geometries. Dust storm control is not so much about abstract control over open horizontal space, as in two-dimensional cartographic renderings of dust storms as flowing lines. It instead works in a complex arrangement of nested three-dimensional zones. Luo's method is fixated first on the geophysics of the near-surface, especially on dune slopes facing into the prevailing wind. There, floating dusts are to be held below the thresholds of altitude and velocity past which they might jump the Helan Mountains and flow beyond the region.

The technique of staged dune stabilization that Luo outlines depends on construction and planting as means of recovering the pluripotency of the wind as a landscaping force. These techniques multiply forestry. If the history of modern state forestry in China (Songster 2003) and elsewhere has been closely linked to matters of the biopolitical comanagement of populations and timber supplies (Kosek 2006, Philip 2004), here, Luo discloses it as a practice of shaping winds to sculpt storm-prone landscapes. This moves beyond trees; shrubs, grasses, and farm waste, not to mention concrete, nylon, petroleum, and plastics; all appear as possible construction materials.

Noting the diversity of materials used in forestry projects, which go beyond trees and also distinctions between biotic and abiotic landscape installations, is a way of emphasizing that these projects are neither about "greening" or the triumph of life over deserts that are often described as lifeless and unproductive. To attempt to shape new landforms is indeed to reprogram potential airforms. Shifting between slides of diagrams and photos from demonstration projects, Luo describes progressive plantings and constructions on a dune's windward face as a sequence of *wind-sand* interventions. Each building on the last, these geomorphing stages progressively reshape the wind and sand in relation to each other. First, nylon or straw windbreaks are erected at the base of a dune to create low patches of wind-sheltered sand, in which well-chosen shrubs or fast-growing trees, native to the place at hand, can be planted. As they grow, they cast shadows of wind shelter higher up the face of the dune, slowly deflecting wind higher and higher up the dune's face.

As plants and barriers redirect the movement of air in a microtuned meteorological dynamic at a dune's surface, higher shrubs or trees like desert willows can be planted as an aeolian infrastructure on a dune's side; these are forms that propagate, wind shadows growing out of wind shadows, until the process nears the dune's crest. "We do not plant all the way up. When we plant high enough on the dune, we stop," Luo says, surprising me. Full coverage is not the goal, tentatively departing from the stated goals of national forestry projects to increase tree cover as an intermediary atmospheric stratum between land and air. When I ask why, he replies that as the bases of dunes are stabilized and wind can be channeled overhead, the wind will flatten the crests of the dunes by blowing away exposed sand above the up-creeping tree line, slowing sand enough that pioneer species can take hold without concerted planting efforts.

A dune's wind-facing side transmutes into a phased construction project, ratcheting up a step-by-step capture of the wind into its earth-shaping mechanics. In a year, the dune's shape will have changed around their installations, allowing them to plant and build yet higher. Each season of construction sets up the next. The staged planting and construction flow chart he describes technically resignify the wind from a free-flowing environmental force into a component part in the shaping of a new geoform.

TEXTURAL MECHANICS

Spreading from the sides of all the new highways across Alxa and near mountain gaps, grids of straw, nylon, and farm waste concentrate winds into jets and unfurl a hypnotic pattern onto the faces of dunes. They tuft the land into the raised grid of a quilt spread over a landscape in process. And yet they nonetheless resolve its shape into microtopographies, which, in aggregate, retain the form of the prior dune. The shape of the large dune remains as a ghostly form (Mathews 2017), cut by wind over the lips of straw grids into terraces of sand. If land form is a history, the gridded constructions casting low dust shadows at the surface are an attempt to shift its course. They work the history and form of the social and economic history of the region as well as ongoing mutations in social governance into the topographical futures of shape-shifting land.

These grids may appear as an attempt to impose a rigid, geometrical order on a world defined by motion, and in doing so attempt to impose a statist order on an irruptive Nature. In fact, they should be understood as part of the technical reconfiguration of a fantasy of technological triumph over Nature into an attention to entangled flows and arrests, the moving parts that compose a moving landscape, and through which it is reshaped. The grids and other sand control installations are not an imposition that masters the landscape, but they are explicitly figured as taking part in its ongoing geo-aerodynamics. This practice of environmental engineering "does not demonstrate that it is impossible to maintain the distinction between work and earth—between technics and nature. Rather, it demonstrates that this distinction cannot have been established in the first place" (Gladstone 2019, 394).

In the proleptic conjuration of a meteorological apparatus, they disclose a sense of the continual and open-ended morphological process that defines the fluid geology of sandy lands. Both the landscapes Xiao describes and the techniques for shaping them converge on the matter of process rather than final formation. These landscapes take shape through "distributed patterns of material structures and the multiple local practices that sustain them" (Mathews 2018, 391). The efficacy of such plantings and constructions is precisely in how they redirect a flow of matter that was already in process. This means that engineering and other technical procedures must not be understood as the mechanical imposition of form on chaotic matter; rather, they represent situated attempts to modulate processes and forces already in play. Engineering a landscape is thus a matter of "interrupting it according to a definite contour" (Simondon 2020, 25). It is not a matter of simply changing the land but of changing how it changes.

In this way, the entanglement of forces, histories, and processes that have revealed places in Beijing's dust shed as cradles of dust storms stand less as conditions to be resolved and more as a set of at-hand constraints that, in effect, can be apprehended as a *de facto* landscape process. Working with these constraints and reprogramming them is not a question of canceling out environmental processes but rather of recircuiting them to other ends. They are not an attempt by an exogenous state attempting to materialize some prior political logic onto an unruly landscape but part of a broader set of tactics that locate political processes and decisions within the geophysics of a landscape defined by its sprawling capacity to change.

Engineers evaluate potential plantings or construction projects as structural interventions into the relation of sand and wind, discerning in the physics of dust a clear relation that makes aerodynamic reengineering of local land a chokepoint for national sky. They approach tree planting as one of a wider arsenal of techniques for blunting and rerouting surface winds, reshaping the properties and textures of the land as a dust-sculpting structure. The spacing of tree walls is calibrated so that their aggregated dust shadows cumulate into an up-scaled weatherproofing, a country's upwinds idealized as files of well-spaced tree walls breaking into the desert. This mobilization of forestry to reengineer material flows has clear parallels with erosion-control measures abroad and at home, as when state forestry programs were mobilized in the late 1990s to stabilize sloping land after catastrophic mudslides in China's south.[10]

To forestry engineers, the value of plantings is described in a geostructural calculus that broadly exceeds calculations of economistic utility or returns (D'Avella 2014, 174–76). Neither is it captured well in reference to state logics of "greening" alone, with their productivist eco-aesthetic preoccupation against lands designated as waste.[11] The tree species I have seen planted for dune stabilization and protection systems are sometimes fruit trees that might supplement local economies as fodder for penned animals or sand products marketed as local specialties (see chapter 3) like desert hawthorns (*shazha*) and other so-called sand commodities (*sha chanye*).[12] But for the most part, forestry officials do not dwell on the economic returns of such plantings done for forestry programs. They choose species for their cost, survivability, and structural properties. They are oriented especially to species that can adapt morphologically to sandy substrate, as when roots that are exposed when dunes move become branches.

Xiao continually reframes the conditions that otherwise drive land degradation or stand as evidence of desertification as potential resources for sand control. This is a practice of technonatural montage in desertifying places and their damaged social and ecological relations. Land degradation and strong wind appear as an embarrassment of local resources that might be reassembled as the constituent parts of new landscapes in dust-storm source areas. Flipping through slide after slide of the plants that thrive in the ruins of wind-scraped pasture, Luo, in a frigid Beijing cubicle, outlines their weedy opportunism as an appealingly efficient choice for mass seedings and plantings, especially for aerial seeding campaigns like the one that opens this chapter. In his estimation, desertification has generated a windfall of sand-loving plants that might be appropriated as the scaffold of a weather-world to come.

Luo's delight over the ingenuity of the approach derives from this continual emphasis on the landscape itself as working, reshaping itself in new patterns and processes catalyzed by staged construction. His is a measured wonder at the magic of the land to continuously change shape. He sees governmental institutions like his become part of this unending transformation, their work registered in a profusion of "sensuous textures" (Gordillo 2014, 7) through which the land and the state refract each other. He collects images of black desert beetles to argue that structural intervention indeed sets off other processes, evidenced in the biodiversity that insects seem to demonstrate. At the same time, he dwells on how such structural interventions, at times, only emphasize the unending topographical change, as when he describes how shrubs that are well-established at the surfaces of dunes might be half-buried in another season as sands move and pile, or how a dune may shift past plants, leaving their root systems in open air like stilts. From his Beijing office, he imagines dust storm abatement as the fine-tuning of a geophysical mechanism that requires only materials, wind, plants, labor, and a bit of patience and flexibility.

I ask him again about the long-term durability of these installed landscapes. He echoes Xiao. "We cannot hope to stop the wind," he replies with controlled exasperation, as if I am too concerned with anticipating their inevitable failure than appreciating their reshaped dynamisms. "What is important is to transform the *dimao*," the qualities and morphology of the land's surface. In doing so, engineering becomes a matter of coordinating disparate efficacies into resonances that might ripple across the landscape to strengthen its wind-stable form by scaffolding wind shadows out of the same forces that cause dunes in general to form. Small dunes reduce and yet echo the flowing dunescapes out of which they are wrought. Out of the dunes, staged planting harnesses the wind to create an undulated forest of cropped peaks, which, in its change, continuously stages the conditions for more changes, each rendered a phase in the landscape's reinvention. The topographic forms that they seek to sculpt out of the open fields of mobile dunes are thus figured as a protean scaffold—at times meant to hold the land still, at others to make it move.

AUTOPILOT

In a meeting of ex-herder men, held in the smoky tent of a tree-planting encampment, Forestry Vice Secretary Zhang fields complaints from ex-herders who have been lost in the often-confusing schedule of payouts for doing forestry work. Antidesertification forestry, from planting trees, to provisioning the seeds and water, to a slew of subsidies for ex-herder participation in state ecological construction projects, has generated its own economies for ex-herder families. Zhang, discussing aerial seeding efforts, argues that dust storms reveal new economic opportunities for families precisely as herding is becoming less tenable, due to land degradation and the government controls on grazing that aim to reverse it. "[Aerial seeding] is also a source of cash for herders [*mumin*] who can collect seeds" from the scrub grasses that spread opportunistically across their degraded pastures "to supplement their incomes." It is one of the replacement economies set up by the local government to feed into antidesertification work in the wake of the dismantling of the region's pastoral economy because it is named as an official cause of desertification.

In this section, I detail how state antidust programs have affected one family, drawing them into the construction, operation, and maintenance of windbreak ecologies and other forestry engineering programs. In the intensification of control over the herding and grazing economies of the region, ex-herder populations have been absorbed into experimental pilot programs that have approached social management as a prerequisite in the transformation of the physical landscape. In the wake of grazing bans designed to stem desertification by controlling so-called overgrazing, forestry programs aim to animate ex-herder housing and economic behavior patterns as essential inputs into dust control infrastructures.

Controls on grazing in the last two decades as a dust-storm mitigation strategy have had the effect of rapidly creating a population of ex-herders who are cut off from the mainstay of the region's economy. They were quickly identified by the local government as a threat to social stability, and, at the same time, a potential labor force available for anti-dust-storm forestry projects. Infrastructure programs depend on experimental and pilot social-governance projects that strand ex-herders in a space of tentative stability that might yet at any moment fall apart. The production of these ambivalent affects, I argue, operates both as a strategy for maintaining a tenuous social stability through holding open hopes even without realization (Miyazaki 2004) while also maintaining a populace that is continually ready to enter into the forestry apparatus, making social governance a key technical component to the operation of antidust infrastructures. In meteorological perspective, a disorienting combination of precarity and opportunity have become materially necessary to both the construction and operation of the kinds of engineering projects described above, and as a key affective technique for ensuring "social stability" by paradoxically maintaining a condition of existential insecurity.

As government investments in the region increase, both through its conversion into a forestry zone for mediating downwind weather through ecological construction and, paradoxically, as a crucial resource frontier for coal, minerals, and wind power installations, social governance has largely taken the shape of subsidies and programs for dismantling the region's pastoral economy as an emergency sand-control measure. While ex-herders are coordinated through strategic economic incentives into new economic ventures, subsidies alone cannot cover living expenses. Nor, importantly, are the baseline subsidies enough for ex-herding families to finance the cars and apartments that have become an obligatory package for their children, male or female, to marry well. This gap drives ex-herders to constantly search for new economic opportunities—most immediately in the new forestry economies of the desert fringe. In this carefully managed neglect, enterprising ex-herders are continually remade as a flexible labor force for sand-control programs; the deterioration of a socialist infrastructure of support (see Berlant 2016) has become a strange mechanism in the construction of Beijing's meteorological infrastructure.

In 2008, Old Tai and his wife, a husband and wife in their fifties, decided to comply with grazing bans, conceived of as a basic antidust measure. The Tai family was one of the first three families in their administrative village (*sumu*), far beyond the leading edge of the Tengger Desert near the internal border with Gansu province, to sell their herd of three hundred sheep and goats and small pack of camels. They moved to a state-built resettlement village for state-designated ecological refugees with Aunty Tai's retired parents and set a room aside for their two adult sons. Getting ahead of the rest of their village, the Tai family had the opportunity to choose a quiet government-issued brick house that was a close walk to a state-provided earthen greenhouse for growing vegetables, which they could sell in the small, county seat of one hundred thousand people. After piecing together subsidies, windfall from the sale of their herd, and money borrowed from relatives and their grandparents' state retirement supports, the family's two grown sons bought trucks and have been run-ning odd jobs associated with forestry, like hauling fencing, water, and building materials for windbreak projects. They travel across the adjacent provinces to work on strings of forestry projects, which keep them away from home for months at a time.

Over the course of a year, the Tais moved into the newly constructed village of similar houses purpose-built to house ex-herding families that the local government calls "ecological refugees." In the following five years, 90 percent of the families in their village moved, some year-round, others on a seasonal basis, lured in by free housing; the promise of new eco-nomic opportunities in the city, many of which have not materialized; and the increasing economic untenability of household herding in the wake of ecological controls on grazing, considered a principle official cause of desertification. Old Tai also cites access to reliable transport and telecom-munications infrastructure, and the proximity to medical services, espe-cially important for his aging parents-in-law, as key factors, not to mention that they considered it highly unlikely that their grown sons would con-tinue on the family's sand-swept pasture as herders. Aunty Tai's brother's family formally complied with bans and swept in early to pick choice hous-ing next door to the Tais. Two other villages have also evacuated the sands and joined the resettlement village, and the three-story administrative building in its center now houses their displaced administrative offices as governments in desert exile. While three villages' worth of people have resettled in these allotments, they continue to retain formal title over the pasturelands they have left and household registration (*hukou*) through the separate village offices installed in the resettlement area.[13]

While local officials argue, and Old Tai agrees, that these new mea-sures, like access to housing without any clarity over whether they will be allowed to stay, should be understood as socialist support and pov-erty relief for families impacted by land degradation, nonetheless, their participation in state anti-dust-storm measures have immersed them in new kinds of precarity. Perpetually framed as a pilot program with con-tinually shifting parameters as an explicit governmental experiment, state decisions over policy, the amount and timing of disbursements, and espe-cially the status of their claim to their new housing and farm plot are in a state of constant vagueness. This open-endedness closely mimics the open-endedness of the landscape processes that relocation programs are meant to kick-start on ungrazed degraded pastureland.

There are few clear administrative time lines or protocols for dealing with grievances, which are handled on a one-off basis by a small staff of enthusiastic fresh university graduates just entering government ranks. The two junior administrators themselves continually lament this administrative confusion and describe much of their day-to-day business as trying to comfort agitated ex-herders and to move their individual cases further along in the black box of local bureaucracy. With no authority to make decisions themselves, they do their best to refer distressed ecological refugees to higher officials, who also cannot and do not make clear promises. The Tais report with resigned exasperation that the local government has continually set and then failed to meet deadlines to settle the matter of whether families have only use rights or more stable ownership rights. In 2017, five years after the first five-year deadline in 2012, the question of ownership is unresolved. In the pilot relocation village, the continual deferral of these decisions has made the last ten years an exercise in state-sponsored uncertainty.

The attempt at stabilization of national meteorological systems has continually generated a sense of existential instability, tempering the family's work to invest in a life off-pasture with the inescapable feeling that they are living in a prelude to something not yet known. The Tais' attempt to offset this by continually building improvements into the house and Aunty Tai's brother's next-door allotment, scraping money together for materials and building on a piece-by-piece basis into a larger multigenerational compound for their extended family. Tai laughs that he is perpetually waiting to buy more cement so that he can literally build a claim to permanence into the house. In fits and starts of frenzied construction and long waiting, they are in effect building up an ethical and evidentiary claim of the family's investment in their own resettlement and, therefore, in the broader state-ecological vision of the region, as his ex-pastureland is slated for forestry windbreak planting.

This form of precarity, I wish again to emphasize, has little to do with the drawback or neoliberalization of the socialist state, as if socialism could be apprehended in a simple scalar measure of responsibility instead of in its continual mutation. Such precarity must be understood as itself a political technology—an experiment in governing rather than the retreat of the state—through specific modes of governmental reinvestment of ex-herder populations and their "social stability" as a technical prereq-uisite for environmental engineering programs, as they move in and out of desertifying pasturelands-cum–windbreak zones, and as they move in and out of the economies that provision, maintain, and tweak the state-engineered meteorological machinery at the land's surface.

The affective consequences of this precarity are a precondition for drawing ex-herder populations into forestry projects; it is the effect of a peculiar mode of governmental reorientation to social life in envi-ronmental perspective. Certainly, contemporary trends in the People's Republic do not indicate a loosening of state involvement in social life but rather a series of shifts in how it approaches the problem. Uncer-tainty, neglect, and unexpected bursts of opportunity that demand quick maneuvering have become de facto modes of affect management, crucial to coordinating ex-herders into provisioning engineering projects with labor and materials.

This condition of existential precarity baked into the operation of for-estry programs keeps families like theirs in a state of constantly frustrated anticipation. As they are working to settle into their new farming lives, decisions over land tenure, the fate of the pilot relocation project, and especially their claim to their houses are deferred. They are perpetually listening for new opportunities, keenly attuned to the erratic stream of side hustles, many of which emanate from the complex series of contrac-tors that have emerged as part of the sprawling economies of state-led dust control. For this reason, resettlement has generated its own lived tempo of punctuated time, structured through an attunement to short-term economic opportunities.

A tense sense of readiness vibrates through their house. During the long days of high summer, after working their greenhouse plot of chives and corn and working on the brick house, they often skip dinner to work with family members and neighbors before nightfall, heading out into the desert and often into newly afforested sandbreaks to collect materi-als to sell in informal markets and often back to state forestry programs. One night they leave in the middle of dinner with flashlights to collect larvae and caterpillars to sell as high-quality-protein animal feed from the exposed beds of disappearing desert lakes. Some days, they take the small truck Old Tai uses to haul vegetables to market into the desert on government roads to gather the mossy *facai*, an auspicious staple of many lunar new year banquets whose harvesting is a staple on the laundry list of official causes of desertification in Alxa.

Forestry officials in the main prefectural town understand these illicit but tacitly allowed trades as pressure valves for maintaining social stability through continual and erratic bursts of economic opportunity. They skeleton a sparse calendar into seasons of semipredictable work, allowing for an affective baseline that orients them to opportunity, sometimes as entrepreneurs and more often as a responsive workforce. Desert tourism ventures on new dunes, for instance, have been widely promoted by local government as a way of framing ecological degradation as a windfall of new opportunities and sometimes perversely as a welcome opportunity for regional economic diversification. When Tai's younger son finally returns to the resettlement village, where he has never lived, and which he cannot be sure will continue to belong to his family, he takes me, with friends and a childhood sweetheart, to race dune buggies over the waves of dunes—the same vehicles that are blamed elsewhere as a cause of desertification. They marvel over how a visiting American could be so bad at driving.

But many more of these ventures are generated by the needs of the forestry and anti-dust-storm infrastructures themselves. Even streamlined, apparently low-labor techniques like airdropping seeds depend on massive inputs of labor and material. Ex-herders like the Tais and especially those who have remained on their ex-pastures collect, sort, and sell the seeds that will rain from airplane holds in aerial seeding campaigns, and ex-herder cooperatives grow saplings that will be used in windbreak projects. In the short, intense springtime planting window just after the thaw, ex-herders stand as a ready pool of quickly recruited seasonal labor as tree planters. Many of these projects take place on the degraded ex-pastures of resettled ex-herders themselves, as much of the region has been designated a dust storm source area.

The room the Tais maintain for their two grown sons, both in their twenties, remains empty for months at a time. In search of opportunity and adventure, the two sons saw grazing bans as a chance to fully invest in new prospects, imagining a life off-pasture. Each borrowed money to buy a truck, much of it disbursed as credit from the forestry administration as seed money for the economies that forestry itself creates. They now work a string of long-term odd jobs as truck drivers, and their main client is none other than the local State Forestry Administration. Ruihong, the older son, hauls water, making endless rounds between deep pump wells and the windbreak projects stretching across the region. Xiaolong, his younger brother, hauls other materials in his flatbed truck, most recently wire for fencing enclosures against the goats that are directly blamed for dust storms. Between jobs, they return to the village for a few days at a time, where they remain on call or seek out new business, listening, like their parents, for word of jobs. While their parents worry over the status of their claim on the house, to their sons, the resettlement village is a crash pad between jobs in the relentless search for work.

As ex-herders seek stability through participation in the sprawling governmental ecology of anti-dust programs, they nonetheless are drawn into new forms of precarity that are crucial in keeping them available as a ready pool of labor, materials, and land. If it appears that pilot programs in antidesertification social management are failing to provide stability, a fine-tuned condition of precarity has become an important part in continually reabsorbing ex-herders into state-led construction of terrestrial and meteorological infrastructures. The ebb and flow of ad hoc opportunities into the lived time of ex-herders responsive to windfall opportunities for work and cash indeed manifests as a replication of the seasonal rhythms of state-built earthworks and of their perpetual need for maintenance.

In governmental perspective, this shift from one form of precarity to another accomplished the dual aim of emptying the desertifying pasturelands in order to reduce dust-storm potential and, at the same time, creating the economic and affective conditions in which ex-herders can be continually drawn back into forestry programs. The flexibility allowed by state pilot anti-dust measures thus at times aims to finely tune social precarity to provision forestry programs with human resources, calibrating uncertainty so that ex-herder families remain available at short notice without, for now, threatening the delicately maintained condition of quietude that local officials call "social stability."

MACHINE SOCIALISM

Back at road's edge, Xiao relights his cigarette. Between drags, he uses the lit end to point out features of the patchy late-spring scrub that spills from road's edge toward the open inside of the land. "It's very hard and very expensive to plant everywhere," he explains, "so we airdrop seeds at key places. Along roads like this, or wherever the wind carries them—where the wind is strongest against the sand." This way, seeds might root where the wind presses most insistently against the earth's wind-shaped surface. Broadcasting seed from planes, Xiao mentions while the driver struggles to light a second cigarette against the breeze, appropriates spring winds into seed distribution systems. The desert is too vast and spreading too quickly, he says, to plant everywhere. It is a technical infeasibility.

Here, he explains, aerial seeding *is* feasible, as if the micrometeorology of rain and wind were designed for it. The mountain range that frames the road funnels rainwater. Strong winds become engineering windfalls: they carry and distribute the seeds just to where the wind is strongest and, so, where windbreaking shrubs are most necessary. A mobile dunescape is a map and a history of the wind's press. A well-aimed pass of an airplane makes strong wind into a windfall, distributing and embedding the seeds just where the wind hits the ground the strongest, just where its power to shape an earth sky is most pronounced.

The bureau buys shrub and grass seeds from ex-herders, who collect them from the sand-loving plants that thrive in degraded pasture. Broadcast from planes, seeds extend fast-growing roots and wild sprays of wind-splitting foliage. Their physiognomy reaches into the touch of sand and wind below and just above the surface, building state hope into the volumetric space where a dune and its wind might coalesce into a cloud of dust. Around each shrub, winds split and slow, to fizzle into walls of wind-sculpted ground or to carve rambling channels into that land, eliciting new physics in its changing structure and shape. Xiao hopes that these channels and walls, revealing what was just underground as new patterns of surfaces and textures, will catch and ground just-forming flows of dust, like silts settling out of a stream in a stretch of slowed water.

The long rise and fall of open dunes and periodic whips of flowing dust is a concert of land and sky, a plastic geophysics that might be resolved into provisional quiet. These shrubs splotch the landscape into shrub-capped mounds, to change its flow like boulders falling into the bed of a swift river. Tracing the wind to divert the wind, these shrubs, grown of seeds collected by ex-herders from ex-pastures and sprayed from repurposed airplanes, are a way that the forestry bureau gathers desertification ecologies and populations, technology, and the wind itself into the protean tangle of substances. Seeding campaigns, like the other environmental-engineering projects in construction across the windy plateau, condense and reassemble Alxa's post-desertification natures and social formations into a campaign to govern the aerodynamics of a mobilized desert and its potential for storm events.

For the bureau, it is a salvo into a terrain. The goal might be finally to steel the land through redundant barriers that assert a provisional separation between solid earth and fluid sky, wresting the surface of the earth into a fixed impermeable barrier. In practice, one must contend with a land that, under relentless wind, shifts between phases, especially in the dust storms that rework geology and meteorology into a continuum of substances that lifts Alxa's land into downwind skies.

Land oscillates between stasis and flow. Geomorphology has become a problem of the land's propensity to periodically undermine its own solidity. Any landscape engineering must contend with this problem, where "land" refers to a dynamic interaction of substances and phase transitions, wrought in the interaction of sand and wind. Xiao's description of the seedings paints landscape as a powerful process rather than a set of fixed properties. It is not simply a matter of dropping a landscape from the sky, as if a borrowed airplane, a borrowed air force, could once and for all fix China's mobile land into a stable grid. Instead, the motion, shift, and change of land and sky, insofar as they have emerged as a problem to be resolved, put environmental engineers in a position not of finding ways of holding the world still but making it move in other ways. In attempting to achieve a modicum of geo-meteorological security, for a plateau that has become a weather problem, they are trying to change how the world changes, sometimes desperately trying to hold it steady, while at other times diverting its relentless flow.

———

Sand is still everywhere. It is in everyone's boots and hair and mouth. It settles into the upholstery of Xiao's borrowed jeep. It falls over the county administrative town, where city workers clad in blue and orange sweep it off the roads and pile it elsewhere, where it blows away again, in a redistribution of matter by the work of people and by wind. On the degraded grassland, giant piles of sand excavated from the road and mine projects that are popping up everywhere heave in the wind as giant man-made dunes, quickly spreading and forming into a new topography. Even at the shrub-secured roadside, sand spills from between roadside shrub plantings to press against the road's sides, tracing out the skid of the multiple traffics that cut across the road. The occasional inundation of transport infrastructures and the inexorable labor involved in clearing them, from Xiao's perspective, are a necessary sacrifice for the benefit of Beijing. It is a quantum of earth kept out of downwind air—a dust storm in negative, collapsed back into land under its own weight in slowed air.

Between weed-capped mounds, trails blaze open in the slalom tracks of motorcycles headed into the land's interior, the traction of slightly deflated tires deepening the ruts of earth-shaping currents. They rush out in desire lines bounding toward the open inside from gaps snipped through the wires of the roadside fence. Footfalls of illegal flocks of goats follow closely in their tracks. They presumably evade forestry police patrolling for clandestine goatherds using the new road to gain access to the foliage of air-dropped shrubs. Officers are well known to accept a goat in exchange for looking the other way. Connecting the dots of pebbly goat scat around and between the shrubs also maps the course of the latest wind in reverse.

Xiao pauses before agreeing that these trespasses might corroborate the topographical process set off by the broadcasted seeds. Family pastoralists insist—and the ex-herder driver, chuckling, concurs—that these flocks may indeed stimulate shrub growth by chewing spent stems and by leaving their scat as a nutritional windfall for the hardscrabble forest. A rogue ecology, ground support for the airdropped seeds. The illicit shit of goats is a nourishing contribution to the reengineering of Beijing's weather in this distant contact zone of land into air. They, too, have been absorbed into the engineering specs of a downwind atmosphere, a machine sky molded into shape at the surfaces where broken land can become bad weather.

In 1924, Sun Yat-Sen spoke amid the rubble of the recently fallen Qing Empire, famously arguing that a Chinese Revolution could not proceed as an attempt to replicate European revolutions. Indeed, Sun, canonized on both sides of the Taiwan Straits as father of modern China, lamented that the goals of a Chinese Revolution must be articulated through the specificity of Chinese conditions. And for this reason, the ends of Chinese revolution must be "just opposite to the aims of the revolutions of Europe." European revolutions, he explained, insist on their own universal character, and in doing so obfuscate the specificity of European history by offering it in the ideal form of an example. His lament provincializes European revolutions by arguing that their proclaiming of "liberty" as a revolutionary political end can only make sense in response to the despotisms that render European rule deeply idiosyncratic—a satisfying reversal of the tropes of oriental despotism that had long characterized European caricatures of China (Hegel 2001, 132–85; see Wittfogel 1957).

For Sun, China's problem was quite different. He explained this difference through what he offered as the specific character of Chinese social forms, each of which he described in a condition of almost constitutional resistance to the organizational forms required of a modernized politicum.

While European revolutions responded to a *lack* of liberty by demanding it, in Sun's China, it was instead a *surfeit* of liberty that was at issue. The over-freedom in Chinese society, he argued, made it prone to disorganization. The realization of a unified national citizenry stumbled over the primordial affiliations of clan and region to which Chinese people would inevitably default. This excess of freedom was a principle of structural decomposing. He argued that this pesky freedom thwarted the consolidation of a scattered populace into a strong Chinese nation, modern and bolstered against foreign incursion.

In his programmatic lament, Sun introduces an enduring metaphor of the Chinese condition. He likens the Chinese populace to a "sheet of loose sand" (*yipan sansha*), whose "particles will slip about without any tendency to cohere." Sandiness, for Sun, is a geosocial condition of "excessive individual freedom," epitomized in the image of Chinese people as atomized particles that elude the formal integration necessary to become a strong, unyielding national body. And it is because, like particles of sand, "the Chinese have too much freedom, China needs a revolution" (1927). In stark contrast to the vital swelling up of nationalistic movements elsewhere, Sun's China is defined by its tendency to perpetual disarray in a metaphor of geological, physical undoing: the un-cohering, and therefore un-capacitated, geological anti-form of sand.

China appears here as a play of sand, society, and state, set in a dynamics of phases and forms of shifting matter. Revolution is a call for a kind of geosocial transformation, one that will furnish the fledgling post-imperial state with a society coherent enough to uphold its claims to political subjectivity. This China, like sand, desperately requires and yet continuously preempts coherence. It invites and refuses stability—and the state—in the same instant. This is the particulate freedom that is the vexing raw material out of which a modern nation must be built. Confronted with such freedom, Sun poses China as an entity requiring an immediate shift in phase, a hardening of loose particles into a solid form still to come. For sand, Sun insists, must "become pressed together into an unyielding body like the firm rock which is formed by the addition of cement to sand" (1927).

The metaphor of China as loose sand, despite its own connotations of shifty, disorganized materials, has become a fixed feature of modern Chinese political tradition, a recurring cliché of a state that unendingly demands stability. In the following decades, Mao Zedong would assert that for the first time in China's history, the Chinese people, under Communist leadership, could "change to being united from being like loose sand, a condition which favored the reactionaries' exploitation and oppression" (1977, 174). Here, and elsewhere in Maoist thought the CCP's legitimacy as a specifically Chinese revolutionary party derives from the consummation of Sun's geosocial vision, confirmed in the execution of the Party's capacity to precipitate a People out of the shiftiness of its constituent particulate populace.

Sand, and the society that takes shape in its image, appears as constitutionally antiformal, resistant and evasive to final organization; for this slippery quality it also seems to invite attempts at its consolidation. Sand moves. Left to its own devices, it breaks apart and eludes any permanent form. In doing so, it makes its own solidity a political end rather than a material given. For Sun, sand, like the condition of a China that could be so easily pieced apart by foreign imperialists, denotes the condition of a specific geologized instability: a powerful passivity, an active inertia, that functions as a structural resistance to the organizational forms required of a strong, modern nation. It disrupts any revolutionary *telos* of freedom into questions of maintaining the specific unfreedom necessary for the establishment of a unique Chinese modernity. This social physics of impending disarray continues, implicit in the continuous calls for the maintenance of social, economic, and political stability that continue to define Chinese politics (X. Zhang 2001, 16).

This conception of Chinese modernity cites European universalisms to mark its difference. Sun puts in play a Chinese political imaginary tilted obliquely in relationship to the liberalisms of Europe and their attendant modernities. This reframing of national emergence as a problem of geosocial shift departs from a remarkable tendency in modern political philosophy to tether political community to life itself, and especially the purposive development of the organism as a key leitmotif for nationalist and anti-imperial politics (Cheah 2003).[1] The analogy of Chinese society with the structural instability of sand, in contrast, names an inertial impulse against the forms of self-organization associated with this organicist conception of nation. It is not a question of cultivating vitalist impulses that already prefigure the nation into being, but instead, of managing the entropic geophysics of society-as-sand.

Sun's geosociology of loose sand locates the state's work in the transformation of one landmass into another, of the constraint of free particles into a stony national structure. To rise, his China must be made to shift between physical states. His lament is an anxiety of power in which the collapse of stone back into sand is only ever held in check. The state thus emerges and operates in a play of substances that slip across forms and phases. Against the tendency of both sand and China to dissolution, the raison d'être of Sun's state lies in the maintenance of a tenuously achieved and only just-maintained cementing. State and society grapple across the tension between the tentative grip on stability and the endless process of entropic un-cohering of sand. The country teeters always at the brink of falling apart again. The national body, like sand, will not, on its own, stay stable for long.

2   Groundwork

ENVIRONMENTAL GOVERNANCE
IN A HOLOPARASITE FRONTIER

Director Luo slams the brakes abruptly in the middle of the empty road, splaying the jeep across its unpainted center.[1] He mutters curses about the camels chewing lazily on the branches of *suosuo*, the endemic saxoul shrub (*Haloxylon ammondendrum*), planted in orderly rows at the road's edge. The bobblehead on the dashboard jolts into its dance with the hard stop. Slamming the jeep into park, Luo pops out of the driver door, looking for bigger stones in the roadside gravel. Without stopping to amass a stockpile he lobs them one by one, in quick and uneven rhythm. They land hard in the fringe of *suosuo* shrubs that spill out from both sides of the road. Alxa's local forestry administration calls their bushy ranks "forest."

Luo, a jovial Han man in his early fifties, once worked as an administrator for an out-of-province coal mining concern. A moment before he had been describing the change of heart that brought him back to the Alxa Plateau where he grew up. For the past two years, he has worked as a local agent and planting foreman for an out-of-province conglomerate based in neighboring Ningxia that specializes in traditional plant medicines, a hot market in an anxious China (L. Zhang 2020). In these two years, his homecoming, his main task has been coordinating the planting of *suosuo* forest on open sand and degraded pasture, for the eventual interplanting of *rou congrong*, or cistanche, a medicinal root that grows underground with suosuo as its parasite. He describes this planting enterprise as a vocation and also as a belated penance to the land after the decades he worked in coal extraction. Solemnly, he offers his work and the dubiously named "forest" as atonement for the *pohuai*, the damage, that the region's coal boom has wrought on Inner Mongolia, which by the early 2000s had become a resource frontier for the mass extraction of the coal that has powered China's economic miracle.

"Damned camels," he murmurs to no one in particular as he scrambles for more stones, which he sends whirring through the whipping branches. He pelts one on the side with a projectile pebble. "It doesn't hurt, their skins are thick," he yells back toward the truck, anticipating my horror. The camel canters away from the road neither quickly nor slowly, an insouciant, middling pace that seems to communicate simultaneous petulance and haughty boredom. It barely deigns to snap to attention at Luo's stone, let alone accommodate the company's planting schedule. The first struck camel draws a few of its herd with it into a stream of bodies, a languorous stampede moving out of Luo's range through the thick of *suosuo* foliage. The flock's movement makes the branches of the *suosuo* rustle and shimmer. Its dust wake spreads into a low haze then diffuses into a cloud that settles onto the sands under the skirt of the quieting shrub "forest."

I look on from inside his truck, bewildered by the dissonance between his description of his environmental atonement and the almost routine animosity with which he pitches stones at the camels. The nearby banner capital touts "The Spirit of the Camel" (*luotuo jingshen*)—hardiness, nobility, tenacity—as one of several new civic branding ventures.[2] It comes complete with a glitzy, cavernous exhibition hall flanked by giant bronze camel statues poised defiantly toward the advancing desert. The spirits of the ex-coal chief, shaken out of the dreamy reverie of his story of environmental penance, turn against the flock of fifteen or so camels, still not running fast enough for his taste. They are chewing the company's nascent forest to cud.

"What a headache!" as he launches stones and obscenities. The unconcerned camels are a carryover of a pastoral economy and ecological formation that is being quickly dismantled by administrative fiat as an emergency antidesertification measure. The central government's mega-forestry project is Convert Pastures to Forests (*Tuimu Huanlin*).[3] Under it, one of the central State Forestry Administrations Six Major Undertakings (*Liu Da Gongcheng*), pastures grazed—and according to local bureaucrats, *over*grazed—by family-held flocks of goats, sheep, donkeys, camels, and horses are to be quickly phased out. Under the program, that land is either to be left fenced and ungrazed, in the hope that grasses will grow back on their own to bind the dunes that have formed after alleged overgrazing, or, they are to be immediately converted into forest cover through the planting of suosuo and other designated grass, shrub, and tree species.

To Director Luo, who oversees the company's participation toward state afforestation objectives, the grazing camels are not simply a nuisance. They are the stubborn ecological fragment (Caple 2017) of a pastoral economy and ecology that refuses to be cleanly consigned to the past. They are a noisy ecological echo in the delicate windbreak forest, a trace of what anthropologist Thomas White calls Alxa's "camel culture" (2016). The camels' eclectic tastes—for both the flora of wrecked pasturage and the fresh leaves of company-planted windbreaks—are a wrench thrown into the medicinal root company's best laid plans. Luo's story of ameliorations, unceremoniously interrupted by the uncouthness of camels, puts personal and landscape remediations in parallel. They mirror each other. The changes in the landscape, with shrubs rising into the dance of *windsand*, seem to confirm the earnestness of his own transformation. They express, in his telling, an underlying narrative teleology of green enlightenment: of open-pit coal mines that become forests, of coal bosses turned ecological Samaritans.

It is, nonetheless, an awkward story. For one, his company's planting site, on either shore of the barb-wired roadside, is on officially degraded grassland: it is ex-pasture, not ex-mine. His company has leased the expanse of mobile, newly exposed dunes, with the local Agriculture and Animal Husbandry Bureau (*Nongmuju*) and forestry administration acting as brokers, from ex-herders who have moved into a newly built resettlement village just outside Alxa's main prefectural town. The renovation of pasture into shrub forest thus finds his work—as an agent of an out-of-province company, no less—closely entwined with state programs for sand control and combating desertification.

Another awkwardness: he leaves out the complex cocktail of forestry policy experiments, market researches, and tax breaks that has lured the company to Alxa, which will be central to this chapter's inquiry. His story of environmental awakening elides mention of the complex new economic environment taking shape to tip corporate and ex-herder behavior alike into the planting and maintenance of windbreak forests. The condition of his personal transformation through *suosuo* is looped into the experimental reorganization of late socialist institutions, logics, and incentives that has made antidesertification forestry a business in the first place. Sand-holding *suosuo* and its medicinal parasite are crucial in these experiments.

The land's apparent repair and the revolution in Luo's conscience are a funhouse mirror; they resemble each other but find no easy correspondences in one another. In reflection, they diverge. Luo's account of his own reform is a stirring story of blossoming environmental subjectivity, in line with analyses in which "technologies of self and power are invoked in the creation of new subjects concerned about the environment" (Agrawal 2005, 166). This framing of new environmental subjects emphasizes the condensation of "environment" into a powerful site of affective labor and attachment. It also centers the cultivation of new environmental subjectivities by state, corporate, and other agencies as a prerequisite for environmental change (Singh 2013). It grounds environmental transformation in the awakening of individuals and communities.

It does not, however, mesh well with either the company's business plan or the strategies of the local forestry administration, whose interests in transforming Alxa's dust-prone landscapes dovetail and split in their respective investments in the hardy sand-fixing shrub. In the staggered schedule of *rou congrong* harvest and *suosuo* planting, Luo sets out a business plan, anticipating state buyers. Profits and reinvestments fall into a rhythm with the machinations of the state, the company's market production schedule, and, of course, the plants themselves.

Luo, pouring himself into the truck after the camel incident, is electric with adrenaline and its aftershocks. The excitement tips the story of his change of heart off its center. In the conversations that occupy us from here into the desert and back, the engine of change has shifted. It has moved from the landscape of his conscience to the economic landscape taking shape with these two plants. He cites not the economy as such (*jingji*), but the economic environment (*jingji huanjing*), an interlocking suite of subsidies, markets, and calculation in which government and conglomerate converge around their holoparasitic botany. Their subterranean inter-rooting patterns into the interdigitation of state and sand as well as the relation between the local government and would-be afforesters like Luo's company and ex-pastoralists turned cistanche farmers. Their symbiosis presages new environments, while consigning others to the disappearances of transition. Luo reminds us of this as he invites *suosuo*'s parasite and laments, with each stone, the tenacity of camels and the inconvenience of their much-vaunted "Spirit."

———

This chapter explores experiments in environmental governance in Alxa through the inter-rooting of two plants: *suosuo*, the saxoul shrub, and *rou congrong*, herba cistanche, a medicinal plant that inter-roots in *suosuo* root structures. *Suosuo* has been identified by Alxa's forestry bureau as an optimal species for controlling *wind-sand* with its quick-growing sand-holding roots and its sprays of windbreaking foliage. The plants' relationship is one of holoparasitism, in which *rou congrong* depends completely on its saxoul host, breaking the surface only to flower and seed. This botanical relation, between a windbreaking species whose planting and growth has become a key goal of local forestry administrations, and its holoparasite, which boosters argue can become a valuable cash crop, is at the center of an experimental reorganization of local government goals and logics, as well as the institutions of late socialist support and poverty relief.

How do the institutional ecologies of late socialist governance tangle into the fortuitous botany of windbreaks and cash crops? And how do the economic and financial conditions of late socialism become part and parcel of governmental experiments that are less about achieving "development" *per se* and more about repatterning financial mechanisms and development institutions, risk, and ideologies of economic behavior *into* the ecological inter-rootings of sand-loving plants? At the center of this chapter's attention are the "biosocial transformations over time" (Masco 2006, 293), unfurling across botanical, bureaucratic, and economic domains. How, against the dynamisms of *wind-sand* in a cradle of dust storms, do late socialist experiments retrofit the organs of rural governance into an experimental political apparatus for making shrubs and roots grow?

Suosuo, thriving in sand and the massive annual variations in temperature of the semiarid Alxa Plateau—from −40°C in the winter to 30°C in the summer—has been identified by the local government as a key sand-fixing species for achieving quotas for installed *mu* of windbreaks and shelterbelts.[4] But more important to the company than *suosuo* itself is another plant, another product. After the two years it takes for the shrubs to properly establish, the seeds of *rou congrong* (*Cistanche deserticola*, En. Herba cistanche, desert-broomrape, Mongolian: *cagan goyoo*), a "ginseng of the desert" (Li et al. 2016, 41; see Wang et al. 2012), can be (im)planted directly into the sand-fixing root networks of mature *suosuo*.[5]

Rou congrong, herba cistanche, is a medicinal plant whose potential usages are attested in the pharmacopeia of traditional Chinese medicine and its modern scientized incarnations (Zhan 2009) as well as in Mongolian ethnobotany in what is today Chinese Inner Mongolia (Khasbagan and Soyolt 2008, 4). Proponents laud its tonifying properties, and recommend it for conditions as diverse as menopause, digestion, and kidney function (C. Z. Zhang et al. 2005). For its suggestive snakelike shape, it was offered to me, tinctured into a potent liqueur, as a supplement for virility, as if by sympathetic magic (Frazer 1966).[6] While no market for cistanche currently exists in Alxa at scale, the local government has promised to create one by guaranteeing it will buy as much as local producers can grow.

Cistanche is a *suosuo* holoparasite. Like other holoparasites, *rou congrong* plants "lack chlorophyll, and they receive fixed carbon, water and minerals from the host plant" (Logan and Stewart 1992, cited in Baskin and Baskin 2014, 869). The holoparasitism of the *rou congrong* radicle is a relation of complete dependence on a host, a developmentally necessary relationship without which it cannot grow or survive. For *rou congrong* and its economy to take root in Alxa, it must *inter*-root. With a suitable saxoul host, herba cistanche can grow indefinitely under the surface of a sand dune, branching out in a tentacular profusion of radicles that twist through spreading *suosuo* roots and archive the shape of a moving dune in its form. It may only break the sand's surface with the searching periscope of a flowering stem, which growers will cover in pantyhose to catch falling seeds before they float into the wind.

Holoparasitic relation offers a botanical form for expressing the relation between two plant species. It has also become a provisional framework for reconfiguring the relationship between forestry and sand control, on the one hand, and economic development on the other. Through the dense patternings by which botanical and institutional forms ripple through one another, I elaborate an experimental modality of social and environmental governance. Here, the deployment of a fortuitous botanical couplet of species by the local government unexpectedly finds the organs of rural rule and social support reconfiguring themselves into a machine for making plants grow into one another.

Two plants shape the experimental reorganization of the institutions and ends of rural governance. As forestry officials try to achieve quotas of afforested windbreak acreage by manipulating the policy and financial conditions for rou congrong cultivation, the relation of a sand-holding root and its medicinal holoparasite reformat the relations of two modes of ecological intervention. I approach the experimental governmental intervention that works in the interplay between multiple modes of botanical, economic, behavioral, and geophysical intervention as "environmental governance." What "environment" denotes here is not a single logic, but, as we will see below, a set of disjunctive logics, arrangements, and political imaginations of process and possibility. Discordant notions of environment gather and interact in the disjunctive botanical relation of *suosuo* (saxoul) and *rou congrong* (cistanche).

Suosuo planting at once achieves forestry bureau quotas for windbreak afforestation and harbors the parasitic herba cistanche, through which forestry bureaucrats and ex-herders turned planters anticipate future cashflow. Planting is thus a kind of groundwork in two heterogeneous senses. The coincidence of a single polysemic term, *groundwork*, mirrors the fortuitous ways in which economic and ecological emergences tangle into the heterogeneous pairing of holoparasitic roots. The multiple senses that meet in *groundwork* thus anticipate the multiplicity of disjunctive, but densely entangling, worldly and planetary considerations that meet in the holoparasitic relation.

Groundwork's first sense captures the forestry bureau's reimagination of its work as direct and indirect action into the physical properties of a shifting earth. The branching roots of the shrub hold the ground below and the rooted branches protect it from the earth-moving winds above. Planting manipulates the geophysical potentials of too-mobile ground. But the holoparasitic planting schemes, wherein *suosuo* is planted so that, two seasons later, *rou congrong* can be grown as a cash crop, is also groundwork in a second sense. It is a kind of preparatory labor, the work that has to get done so that work can get done. For out-of-province conglomerates and ex-herders to become cistanche growers, they must first carefully plant and maintain their windbreak host plants. The market for medicinal roots requires the windbreak forest, and therefore the parasite is required for its host. The earth-stabilizing work of planting thus becomes an anticipatory work for the "real" economic activity. One kind of groundwork conditions the other.

Incompatible pairs shape both political experiment and anthropological sensibility. Cascades of paired botanical and institutional forms ripple through the changing geophysics of Alxa's sands. In the disparate and simultaneous senses of "groundwork," for instance, botanical relations spin out into political formations that hold together the strategic transformation of both ecological and economic environments. The inter-rootings of plants and the realignment of state ecological goals and economic techniques begin to resemble each other. Inter-rooting, as a governmental principle, forces us to attend to assemblages of different notions of "environment" while eschewing the impulse to resolve into a single seamless political logic. The relations between two plants twist, fold, and stretch late socialism into an experimental topology of power: a "configurational principle that determines how heterogeneous elements—techniques, institutional arrangements, material forms and other technologies of power—are taken up and recombined" (Collier 2009, 89).

The modes of ecological governance that follow thus all require relations across difference. Ecological environments and economic environments operate in relation to one another. But they cannot be reduced to any single logic of "environment" that would obviate the differences between the consequential polysemy of "environment" itself, in both English and Chinese. In this, they are like the plants themselves: windbreaks and their parasites do not form a seamless mega-organism. Their relationship encompasses the unresolvable difference between them. Holoparasitism: a scaffold for the relationship between two disparate senses of "environment."

Two Environments

Local forestry bureaus are bound to quotas for afforested acreage, requirements that are attached to the central forestry mega-project Convert
Pastures to Forests. This program builds on a longer history of state-led
ecological construction programs, proceeding as a suite of social and environmental interventions gathered under the ecological rationalist rubric
of the "greening of Western China" (Yeh 2005). In recent decades, as an
explicit corrective to shelterbelts planted with exogenous trees, which
have largely died in harsh local conditions, Alxa's forestry officials promote *suosuo* as a botanical vehicle for hitting the program's substantial
afforestation targets.[7]

The botany of medicinal roots and the forestry shrubs they require tangle a wide array of state institutions and strategies into new sandscapes
under the rubric of *sha chanye*. *Sha chanye* are so-called "sand products"
(see Qian 2012) that thrive in the sandy conditions associated with desertification and other kinds of land degradation. The local government has
been roundly promoting select sand products as successor economies for
an ex-pastoral Alxa. *Rou congrong* in particular is attractive for forestry
officials. They reckon that the promise of a market for *rou congrong* can
move ex-herders, still reeling from the emergency antidesertification measures that have intensified controls on grazing, to convert their former
pastures through *suosuo*-planting. Forestry officials intimate that promoting sand products is a more powerful motivation for would-be growers
to plant and maintain windbreaks than even direct forestry subsidies for
planting *suosuo* on its own. They count these shrub forests, grown as an
ecological prerequisite to cistanche, into their regional afforestation quotas. And the institutions of a redistributive socialism are repurposed by
officials into a set of increasingly improbable mechanisms for manipulating markets that will guarantee a state buyer for *rou congrong*, even in the
absence of any actual consumer demand for it.

Holoparasitism thus becomes "a lively zone of embodied connection
and friction" (Ahuja 2016, ix) in which the extant conditions of rural governance are to be revisioned into an apparatus in vital service of windbreaks and their parasites. Their inter-rooting offers, for local cadres, a
profoundly physical template in the potential intertangling of markets,

botany, and forestry, and the multiple notions of "environment" that circulate in these disparate domains. As anthropologist Lisa Hoffman reminds, the term "environment," *huanjing* in Chinese, is polysemous. It refers to more-than-human nature, but also, more abstractly, to an arrangement of things. An environment is a milieu in and out of which things can be made to happen. For Hoffman, it is most broadly the set of "conditions that make something possible" (Hoffman 2011, 107). It is also a formal principle that doubles as a theory of happening, set in the mutual interaction elements whose arrangement might shape or even determine specific outcomes and effects: a set of conditions in which something is held, and is open to rearrangement.

In this craggy terrain of environmental governance, "environment" works through a parallel movement of differentiation and coordination. It is precisely the doubling of environmental meanings that drive cadres to seize on these two roots to think about economic and ecological transformation together. As their botanical relation has offered a formal principle for the disjunctive integration of state forestry and economic development techniques, their biology is also becoming increasingly integrated into a complex array of financial instruments and the state promotion of speculative markets for medicinal roots. Where heterogeneities scrape and slide into relations of "generative friction, generative enfolding" (Haraway 2016, 61), singular logics of governing give way to a complicated coordination of near misses and double meanings. They are a heterogeneous pairing, "an inclusion that would not obliterate separateness" (Strathern 1988, xv). They articulate the disjunctive and yet conjoined senses of environment together as a template for political calculation and strategy. Local cadres, as they replumb the habits, institutions, and sensibilities of late socialist governance into the holoparasitic fold of roots, increasingly find the weave of the plants' roots also weaving into the changing architectures of their institutional ecologies.[8] Like the officials and planters who populate these pages, I do not approach holoparasitism as a moralized relation between host and its vampiric stowaway, but instead as a format of botanical relation that powerfully shapes how other relations and connections might take shape.

Where statecraft is a radicle emergence, governance works through strategic mutation diagrammed through botanical relationships. Late socialism, with roots, appears as an experimental sympoetic practice, one in which governance and the transformation of more-than-human ecological and economic worlds are constantly becoming-with plants (Goldstein 2019, Mettrie 1996). By focusing on experiments in the coordination of divergent modes of ecological governance *through* roots, I show that even as entrepreneurialism, markets, and finance schemes are absolutely crucial in governmental experiments, any straightforward reduction of these governmental experiments as a story of transition of socialism to a preordained neoliberal capitalism fail to capture what is at hand.[9]

In the next section, I detail the modulation of this economic environment more closely to show how techniques of economic development are recircuited in two ways: that socialist institutions are being recircuited into venture microfinance agencies for would-be root farmers; and that ex-herder behavior and motivation are increasingly being understood, by forestry and other officials, as elicitable effects of changes in the economy, imagined as an environment, or *huanjing*. In the concluding section, I return to questions of environmental subjectivity and agency in relation to the inter-rootings that are reshaping land and labor across the windy plateau.

Two Triangles

In the early 2000s, as a string of major dust storms pummeled Beijing, a new meteorological villain splashed onto the city's political scene. In 2000, then-Premier Zhu Rongji toured Alxa, a region coming into infamy as a cradle of the capital's bad weather. He attributed the region's desertification, and by extension, Beijing dust storm woes, officially, to a new problem: over grazing.[10] As opposed to earlier official explanations of desertification, which overwhelmingly blamed the backwardness of pastoral peoples and, in Alxa, the region's ethnic Mongolian population (Williams 2002, Bulag 2002), twenty-first-century panic about overgrazing *qua* engine of bad weather expressed itself in public debate in images of teeming animal excess. Images of herds of sheep and goats presented grazing animals as a ravenous engine of desertification, breaking land and releasing dust into capital-bound airways. Herders were identified as the locus of the problem. But instead of a question of civilization backwardness or environmental irrationality, herding was quickly becoming a problem of contradictory economic and ecological demands. The herder, that is, shifted from a backward political subject to a problematic assemblage at the more-than-human entanglement of animal, plant, geophysical, and political economic forces and temporalities.

Zhu argued that drastic action must be taken in Alxa as a first step in transforming weather in the downwind capital. State-run newspapers circulated his demand as a slogan-triplet: "Kill goats, protect grass, and protect Beijing" (*sha yang, hu cao, bao Beijing*) (*Renminbao* 2000). In Alxa, a decade later, this line was repeatedly cited by Alxa officials as a pithy summation of an approach to emergency ecological construction, pegged to the killing of goats that were the mainstay of the region's pastoral economy.[11] The slogan puts forth a succinct theory of ecology, geophysical structure, and meteorological intervention. It reiterates a sense of Beijing as a city threatened by upwind mismanagement, locating that threat squarely in the interspecies relations of plants, animals, and, implicitly, municipal and regional administration.

These concerns crystallized in campaigns to resolve "four overs," official drivers of desertification and dust formation phrased in the language of socialist mobilization: overcutting of fuel, over-farming currently used land, over-reclamation of *new* land for farming. The fourth over, overgrazing, occupied the center of state-led land use change rather than, say the region's rapacious licit coal and rare earths mining (Klinger 2017).

Without realizing the massive culling of the region's stock that Zhu called for, officials in Alxa implemented controls on grazing soon thereafter as an emergency ecological measure for the protection of dwindling grasses and, by extension, Beijing air. In the early 2000s, mandatory partial grazing bans, which ranged from seasonal cessations of open grazing to outright enclosure of heavily degraded pastures, were piloted in heavily desertified parts of the Alxa Plateau and in the immediate outskirts of Bayanhot, the region's main city. While animals like sheep and goats could still be raised, they could not be grazed on pasture for large sections of the year. To comply with bans, herders were either to sell their herds, or to raise them over much of the year in enclosed paddocks (*juanyang*), fed with expensive purchased fodder. Bans and grazing controls soon rolled across the Plateau as a wave, reorienting herders' relationships to their land and to their flocks in anticipation for the formal end of the pastoral economy as a sand-holding measure. Bans and other controls on grazing continue on as a durable feature of a new regulatory environment for the control of activity on pastures, now each susceptible to the eco-economic scourge of overgrazing.

In Alxa, measures to control overgrazing did not target herders *per se*, an essentialized population. It focused rather on herd*ing*, now reframed as practice that—like the suosuo and rou congrong that would come to replace it—appeared as a meeting point of ecological and economic environmental condition. That is, "overgrazing" was described by forestry officials I spoke to not as a backward cultural predilection of ethnic Mongols, but as an economic behavior that was itself the predictable ecological effect of the maladjusted, over-free economic environments of early Reform. Overgrazing, for forestry officials, appeared as a regrettable but, in retrospect, expectable consequence of new social, land-tenure, and economic conditions following land reform in the 1980s. Communal pastoral production brigades, composed of multiple families raising mixed herds on giant pastures, were divided into much smaller family plots which could not support seasonal shifting of flocks among different pastures (*youmu*), leading to *de facto* sedentarization on smaller pastures which were intensively grazed year-round.

The deregulation of the planned economy in pastoral products began in the 1990s. Newly open to rapacious market pressures after the shelter of the planned economy, ex-herders described themselves suddenly thrown deep into the ocean of business and competition, and subject to the boom-and-bust markets in meat, wool, and especially cashmere—"soft gold" (*ruan huangjin*) in local parlance. These products are crucial to the material cultures of affluent urban Chinese life, and for new international markets, looping newly de-collectivized herding families quickly into expanding networks for commodity animal products.[12] Deregulation and land reform were modeled, in a pastoral context, on the decentralization of farming in China. NGOs and forestry officials alike explicitly discuss Reform Era deregulation of markets in animal products in terms of overburdened pastures and questions of carrying capacity (Sayre 2008).

In their reconstruction of the drivers of overgrazing, forestry officials point to new markets, new pressures, and new spatial and ecological conditions as part of a new context that more or less guaranteed "irrational land use." At the center of this conception of conjoined economic and ecological change, is a conception of herders themselves as quasi-automatically reactive to signals in what, like Luo, Foreman Zhao of the forestry bureau calls an "economic environment." This environment, as a set of conditions that make something possible, was not yet but could and must be manipulated by the state in order to direct a wholesale change in economic behaviors, and then, in the physical environment.

This sense of a reactive environmental subjectivity is tinged with a naturalized sense of a survivalist instinct that could, with proper signals, be harnessed to new ends. This notion of environmental reaction is, to forestry officials, not an essentialized shortcoming of ex-herders, lacking proper care or attention for "the environment," and so is not fundamentally a matter of incomplete subject formation. Nor, for the forestry bureau, is it a question of the moral failings of gluttonous goats, who, after all, must eat. Rather, it was understood as a question of an improperly calibrated set of economic conditions that, in effect, caused overgrazing and its attendant environmental degradation. The question is one of herders reacting to an economy, figured as an environment, out of sync with ecological processes and, finally, state ecological construction goals. The problem exists in the environment, the economic conditions of human behavior in the Reform era.

Ex-herders often echo this explanation of overgrazing as a practice stoked by a deregulated market in animal products. Mr. Jin, a Han ex-herder in 2011 was deliberating over whether or not to comply with grazing bans.[13] He recalled, "When we saw the price that cashmere was getting, what could we do but raise as many goats as we could? In those years, people used to call cashmere 'soft gold' for the price it could fetch on the market. Much, much more than sheep [wool and meat]. So for many years we had goats, more every year. This happened during many years of drought, so each year the sand spread." The Jin family elected in 2013 to move off pasture to a relocation village. Altan, a Mongolian herder, expressed similar feelings of inability to reduce flock size. He reported that, as the high price for cashmere remained high even as grazing bans extended, his family was caught between the choice of going into debt purchasing fodder to feed penned goats or selling the flock entirely. They opted to remain on their pasture, and collect state retirement benefits, only raising a few goats, illegally grazed and easily hidden from forestry police, for sentimental reasons and family use.

Himself from a Han herding family, Foreman Zhao frames overgrazing as a clear reaction to the unregulated market in cashmere. It is an interspecies behavioral signature of Alxa's Reform-era economy, playing out as a maladjusted relation between herding families, the size and composition of their flocks, and the sand-binding flora on which they feed. "When people realized how much cashmere could fetch, their herds grew, sometimes by tenfold in just a few years. The government promoted cashmere then, too, because it was a way of achieving local economic development, and so everyone got goats. These goats cause environmental damage (*huanjing pohuai*) because they eat everything, even the roots of grasses, and they use their hooves to break up the ground. But people had no way not to raise (*meifa buyang*) goats because of the high price of their wool." Local forestry official accounts of "*ir*rational land use" thus understand it as a *rational* response to the new economic conditions of socialist marketization, fallen too quickly out of state control through deregulation.

The notion of overgrazing, in Foreman Zhao's forestry bureau thinking, is a question of contacts, entanglements, and misdirected behavior, all drawn together in a desertification opera of goats, grass, and Beijing. He understands desertification as an effect of overgrazing, which is itself the inevitable behavioral outcome of a poorly controlled economic environment. Human behavior, animal appetites, and whether or not grass can hold the earth, are recircuited into the play of economic, ecological, geophysical, and meteorological conditions. Together, they are a problem of the relationship between the different "environments" through which a landscape takes shape. Zhao's administration's explanation of overgrazing approaches it simultaneously as the economic behavior of herding families who do not act as much as they *cannot not* act; a maladjusted interspecies assemblage; and an effect of specific configurations of markets and social supports.

Note the formal resonances the ecological-economic triangles of dwindling ground cover, commodified grazing animals, and herders, on the one hand. And on the other, the landscapes that local cadres have proposed to replace them: sand-binding *suosuo*, potentially valuable *rou congrong*, and an expanding population of ex-herders. Following Zhao's *post hoc* account of the causes and consequences of overgrazing, the grazing bans that were and are continuing to be introduced across Alxa should certainly be understood as a mode of "green governmentality." The bans make "local herders visible and accountable for their purportedly degradation-inducing range practices" (Yeh 2005, 24), ultimately aimed at the dismantling of pastoral lifeways in the service of a national meteorological security centered, emphatically, two days' dust-flight downwind.

While grazing bans certainly operated to dismantle the local pastoral economy, they must be understood productively as a political technology for creating the conditions of an ecologically adjusted, post-pastoral economic environment, especially through financial mechanisms. Compliance with bans, explained families who opted in, was not only a sudden shock to the embattled pastoral economy, but also a mode of enmeshing oneself in a new economic environment largely shaped by local officials and governmental institutions. Bans must thus be understood as a state strategy that conducts herding families from one set of economic conditions to another. It is an administrative gateway to a suite of subsidies, payouts, and financial and political benefits. Many ex-herding families opted in, anticipating that at some point bans might become obligatory anyway, and perhaps without attached monetary benefits.

On the part of forestry officials, the state-led calibration of the timing, amount, and conditions of state moneys and supports was important as a means of using economic and other mechanisms to choreograph the transition of ex-herders from pastoral production to livelihood pursuits amenable to state goals and sensibilities. Foreman Zhao is careful to note that these benefits strike a sweet spot. They are generous enough for bare living expenses in the short term but not so much to sustain an especially comfortable life, and definitely not enough for important expenditures like throwing weddings for grown children. They are not quite enough. This was a point that was raised by every ex-herder I met.

Promises of access to these forthcoming opportunities, Zhao offers, allow for immediate reemployment to blunt the much-feared social instability in the sudden creation of a population of herders with neither herds nor pastures: a symbiotic coordination, through roots and their markets, of forestry, economic development, and late socialist worries over looming challenges to social stability, especially in a minority region. Zhao states,

"There will be a certain number of people who will take the payouts to live off, like old herders who will not adapt to new ways. But most of them will take that money and use it to start new businesses." This entrepreneurial sensibility, framed as a naturalistic reaction to economic conditions, could be nurtured with promises of state-backed marketing campaigns, preferential and expedited loans, and permitting and funneled into a number of new sectors. These preferred sectors, at a minimum, align with local development goals especially in tourism, services, and rural development. At best, they could reinvest ex-herders on their former pastures, now as afforesters, animated by the potential payoff in specialty desert products like *rou congrong*.

Forestry officials wielded grazing bans as a means of collapsing pastoral livelihoods, which they offered as a necessary measure in a crisis of desertification that was playing out in dust events that roved through the national topography like atmospheric riverbeds. They sought to foment shock at what, by the early 2000s, was widely posed as the maladjusted relation between the pressures of a deregulated market in pastoral products and dwindling grass cover, playing out in dust storms over Beijing.[14] Here, however, local officials, from the low-level bureaucrats in training who were often tasked with fielding my questions to the party secretary, continually described the overgrazing that grazing bans were meant to control as a problem of economic behaviors brought on by unregulated markets that were introduced too haphazardly in the broad reforms of the 1980s and '90s.

Socialist marketization, offers the secretary, was an education in understanding herder behaviors in new economic conditions, set against demands for local officials to bring economic development to rural western regions of China. The behaviors of herders appeared on the one hand as an effect of changing economic conditions, which could be systematically transformed through strategic government intervention. On the other hand, these behaviors, framed as a range of effects of new economic conditions, were also to be understood in terms of their effects on the geophysical properties of degraded pastures and new desert lands, lurching into dust events in conditions of water shortage and aeolian desertification. The manipulation of human behavior, in the understanding of economic and ecological governance required by the central government's demands to "kill goats, protect grass, and protect Beijing," poses that behavior as a seam where two different conceptions of environment meet. This theory of behavior, operationalized in experimental afforestation efforts that depend on creating markets for medicinal roots, fortuitously replicates the holoparasitic relation of *suosuo* and *rou congrong*. Each is a place where economy and ecology touch.

Mandatory partial controls on grazing, piloted in the early 2000s, eventually intensified when it was not simply excessive herding that must be controlled but the economic environment that "caused" it. If high demand for cashmere drove over grazing, then the state aimed to dismantle the pastoral economy by dismantling supply. If grazing bans induced shock, however, they were not aimed at creating new conditions of free market neoliberalism, marked by the reinvention of the state as a guarantor of the market as an ultimate distributor of goods and the good. Instead, they were offered by local officials as an administrative gateway that would channel herding families and their degraded pasturelands from one set of economic conditions to another—they were aimed, that is, at shaping a pathway for ex-herders to move from an over-free market into one that was highly state-managed and oriented toward a suite of local development and afforestation goals and principles.

Bans presaged the experimentally managed revival of a quasi-command economy now reframed as a set of financial and market institutions. Cumulatively, forestry officials argue, these institutions allowed them to pragmatically manipulate the immediate affective, material, and economic environments to which ex-herder decisions were understood to react. Compliance with bans gave newly minted ex-herders access to a raft of financial, political, and technical benefits, oriented toward providing state support for new entrepreneurial ventures in sectors identified by local officials as the anchors of a post-pastoral economy. Compliance in grazing bans would unlock access to new streams of micro-venture capital, compiled by a budgetary patchwork from existing institutions of socialist redistribution and support, as we detail below.

But, here, I emphasize that in rural China, socialist marketization is not so much a process of the freeing of markets and the devolution of state care into an injunction to induce ex-herder entrepreneurial initiative. Instead, it has involved a retooling of socialist institutions and state-controlled markets into a mechanism for converting landscapes, populations, and ecologies from goat-grazed pastures to plantation duocultures of windbreaks and medicinal parasites.

Grazing bans, as we will explore in the next section, sought to create a post-pastoral economic environment, generated and conditioned through intense state intervention, and enacting new ideas of social behavior as a product of well-managed environmental conditions. This transition of pastoralists into tree planters is an enmeshment of late socialist governance into the complex rootworks of a sympoietic state-led landscape and market transition. As we will see, compliance with bans is not a simple expression of a will to decide. They fall outside the easy binaries of the voluntary and the coerced that are often associated with authoritarian rule in China. What local officials offer to herders as a free choice instead appears, for those herders, as a response to the relative prospects of different choices, made in an environment that has been configured and massaged toward specific outcomes.

Bankrolling Symbiosis

March 2012, Alxa Plateau. I've tagged along with Mr. Li in his jeep, as he rides the high dunes that were once his family's pasture. He's gotten lost and he stops multiple times to radio the leader of his hired planting team by walkie-talkie. The dunes are so high that we cannot see over each ridge, making the vastness of his family's former pastureland feel both infinite and claustrophobic. The just-thawed earth sprays loudly against the metal doors of the car. The wind makes the old jeep shake even more.

Li assures me with gruff good humor that he is sure the forest is here somewhere. Garbled messages crackle through the walkie-talkie. He grunts affirmatively. A pause to get his bearings before he punches the ignition, sending the jeep barreling down the windward face of a high dune. When he sees the red flag and faint silhouette of water trucks (fig. 5), he hits the gas gleefully. The corner of his pasture is now a field of sand drifts.

Two years earlier, in 2009, his family, facing the dual economic pressures of desertification and the constraints on grazing implemented to control it, decided to comply with full state bans on grazing, selling 90 percent of their herd of two hundred goats. The land, risen into giant dunes, is again

Figure 5. Ecological construction, Alxa Plateau, April 2012. Photo by author.

buzzing with activity. His hired team is planting and watering *suosuo* for the second time today, after the first batch was uprooted by the hard spring wind. The team is busy replanting shrubs in straight rows that trace a fluctuating grid on the rising curve of the dunes. They dress in military fatigues and red armbands. This is the typical uniform for patriotic tree planting campaigns. However, it is their former neighbor, Li, who has hired them for the week. Bright scarves over their faces protect them from the dust that whips around our boots and peels off the surface of the sand.

I walk with him as he inspects the team, mostly ex-herder women hired as day laborers, moving in groups of three. It is the short planting window just after the early spring thaw, when *suosuo* has the best chance and most time to establish properly in rows before the harsh winter. They plant files of *suosuo* saplings in straight lines across the undulating topography of sand dunes, one sinking a shovel into the loose sand, one carrying a bundle of saplings, a third setting them upright in the ground. Moving slowly between these rows, ex-herder men drive trucks hauling water pumped and purchased from a neighbor's deep well. They spray and tamp down wet sand around the scraggly root balls of the saplings, each the width of a finger, listing convulsively in the unrelenting wind (fig. 6).

Figure 6. Groundwork. Note a thin *suosuo* sapling in the left foreground. Photo by author.

From a tentative promontory on the windward edge of a dune, Mr. Li invites me to imagine this expanse of sand as a future forest of shrubs, dampening the hard wind into a pleasant breeze, and catching the wayward dust in its brambly foliage. He envisages the dunes, now convulsing with dust thick enough to make the water truck disappear from view, as a picturesque and productive landscape of *suosuo* gridded in straight ranks. Below ground, the shrub forest will gestate a living mine in *rou congrong*. He describes these plants-to-come as a way out of the crisis of desertification and grazing bans, both of which conspired to force the family into giving up the herd. Because of state backing for *rou congrong*, the land, he hopes, will return to productivity. He looks forward to that return in anticipation of a return on the family's substantial investments in time, cash, and worry.

However, solvency must wait: it is not until the shrubs are fully established that the work of planting *rou congrong* can begin. And it is not until *rou congrong* matures that it can be sold. And it is not until the local government formalizes its long promise to buy *rou congrong* produced on ex-pastureland that he will feel stable. In the meantime, he is concerned with amassing enough money to fund the economic and ecological transformation of the land. He frets over balancing new sources of investment capital against the costs associated with retrofitting sand dunes into the basic infrastructure for a root plantation.

He runs down the columns of a mental spreadsheet arcing toward the black in the mathematical play of rising costs and anticipated returns, of investment capital and the multiple streams of funding from state institutions, which he figures as somewhere between redistribution and venture capital. *Suosuo* saplings come cheap, highly subsidized by the local forestry bureau, with payouts rewarded for forest survival by *mu* for the first two years before they establish. Then he has to hire and feed the worker teams who plant *suosuo*. Funds disbursed by Alxa's forestry administration and other offices of the local government are part of the mental ledger through which he understands the coming years in a temporality shaped by plant and subsidy cycles.

These subsidies are funds made available through savvy negotiation of state forestry as a new kind of economy. The family made a large lump sum from selling the goats in one fell swoop. This pools with the family's savings, loans from relatives, and the retirement disbursements of his and his wife's surviving parents. Payouts for first planting of *suosuo* and additional yearly payouts for acreage of surviving *suosuo* round out his budget. For his family's voluntary participation in grazing bans, he receives a yearly sum from Alxa's forestry administration, tracked to the total area of the ex-pasture. Supplementary poverty relief funds (*pinkun jin*) for each person whose household registration (*hukou*) is logged on the family's land are also added to the pool. Though his adult son lives off-pasture, he continues to claim his household registration in order to increase the number of support-eligible registrants in the household and on the land.

Li allows us to understand how the landscapes of late-socialist rural governance are being reconfigured as a changing terrain of opportunities and supports oriented around the growing conditions of state-sponsored botanies to come. He approaches the institutions of socialist redistribution as a dispersed public-private financial agency. A multitude of state institutions, including but not limited to the forestry bureau, have increasingly appeared to ex-herding families like his as sprawling regional investment banks, each with idiosyncratic conditions for unlocking moneys that he thinks of as credit and capital for his duoculture of shrubs and roots. According to his business plan, multiple pools of state support will sustain his family in the two years that *suosuo* requires before it can be interplanted with the medicinal root, tiding them through this initial period of groundwork.

Supply Chain Ecologies, Economic Niches

The timing is right, he assures me, noting the medicinal herb companies that the local government has partnered with to set up shop in Alxa. Cutting to the chase, he explains that among other things, these companies are a state-supported guaranteed buyer for his roots, which makes his partially subsidized bid to grow *rou congrong* a difficult, but very low-risk investment. "When it is ready for market, there are processing centers in town we will sell to. They will buy as much as we can grow." Proceeds from sales will roll back into his *suosuo* planting fund so he can expand production, setting in motion the ex-pasture as a self-sustaining machine of step-by-step ecological and economic change. More *suosuo* means more potential *rou congrong*, which, in the state-guaranteed purpose-built market for roots, means more income. After the end of grazing and the soft gold of goats, Mr. Li sees in the grid of groaning saplings the materialization of a new livelihood. Installed into a state-built economic environment, it will, he hopes, spin sand into gold.

For prospective root growers like Li and Luo's company as well as the local cadres who are trying to channel their energies into state forestry goals, markets for roots are crucially important to their understanding of their own action. There is a "simultaneous redefinition of the environment and the subject" (Agrawal 2005, 23–24). However, the mode of environmental governance at hand has depended on markets as a way of intensifying state intervention, not reducing it, especially when it is precisely the over-freedom of markets that local officials cite as a driver of overgrazing. Markets, instead, are for local cadres part of the whole financial terrain in which economic behaviors and decisions take shape. Manipulating them properly is at once an important means of governing both social and economic behavior. In an economic environment reverse-engineered for *rou congrong* from start to finish, from the subsidies that fund its sand-fixing infrastructure of inter-roots to markets lying in wait, we witness a holoparasitic reversal: a parasite has come to anticipate its host.

In this section, I focus on state manipulation of demand for *rou congrong*, on governance as a meticulous and experimental attunement to the modulation of the economic conditions that local cadres present as the immediate context for human behavior and for the achievement of environmental goals.[15] In doing so I center a specific modality of environmental governance. It begins in grazing bans as a mechanism of ex-herder behavior modification through payouts and subsidies, as detailed above. And it ends—and restarts—in the promise of expansive and stable markets for the ultimate product. Together, these technologies of intervention into the economic environments structured around the holoparasitic plant pair also attempt to close human behavior into what I call a supply chain ecology. Crucial in this is a specific sense of an economic milieu as an efficient governmental environment through which desired outcomes can be elicited indirectly, through well-calibrated modification of the environments in which ex-herders live.

Where grazing bans created the conditions of new economic endeavors for ex-herders, local officials maintain that the creation of new markets was necessary for their immediate reabsorption in an economic milieu, amenable to state modification, which was attractive enough to draw ex-herders into the play of its forces. This conviction was based on multiple enthusiastic promises by local officials. In addition to root-processing companies like Director Luo's that have been coaxed to Alxa through tax breaks and promises of government goodwill, representatives of the local government have promised that the banner government will buy any locally produced *rou congrong*.

Rou congrong, as a potential commodity, sits between two cachets. First, a burgeoning health market around products and practices that support a craze for *yangsheng*, life-nurturing practices and vital nourishments (Farquhar and Zhang 2012) amid the anxieties of contemporary Chinese life. Relatedly, non-Han "Chinese" ethnomedical traditions, in a complex play of internal orientalisms (Schein 1997), has made ethnomedicinal branding attractive for a coming market, and especially attractive to local officials as one of several products that can be marketed as a local place-branded specialty product (*tutechan*).[16] However, as government officials, company representatives, and growers alike were quick to note, no such market for cistanche currently exists at scale, and certainly not one robust enough to dampen the risk they would assume in an open market. The maintenance of artificial demand, and therefore a guaranteed buyer that would eliminate economic risk from ex-herder decision-making, was understood by forestry officials as the necessary closure of an economic loop.[17]

The economy here appears as a relevant context for the manipulation of human behavior. It has shifted into a governmental mode of environmental psychology. Centering human decision-making and behavior as systematic reactions to environmental conditions rather than as the expression of an individualized human subject, it resonates with Foucault's discussion of "environmental technologies," associated with what he calls American neo-liberalism. In his reading of a proposed measure to control drug use in the US War on Drugs, he reads closely the artificial manipulation of drug prices as a way of manipulating the market milieu in which drug users operate. The manipulation of prices—the immediate and efficient environment for drug users, framed through the psychology of consumption—is the template of a kind of environmental intervention that acts not on the individual user as such, but "on the market milieu in which the individual makes his supply of crime and encounters a positive or negative demand" (Foucault 2010, 259). This depends on markets functioning not as a domain of disembodied collective will, an invisible hand, but rather, as one end of an environmental configuration—the economy—subject to minute and ad hoc modulations of economic stimuli.

What holds together these "environmental technologies" is that the economy reconfigures into a set of relevant environmental factors through which human behavior can be indirectly redirected. Individuals and populations are not figured as environmental subjects, in the sense that they must be made to act with some kind of environmental consciousness—an individualized sense of responsibility that has become crucial to liberal capitalist environmentalisms. They are environmental in the sense that subjectivity is increasingly understood not as a problem of individual agency in an open market, but as the systemic response to conditions that can and must be manipulated by state agencies. The market becomes a terrain for the elicitation of a range of different behaviors, like the apparent decisions to comply with grazing bans and to plant state-preferred desert-holding plant species.[18] The state, here, appears as an agent that seizes on naturalized proclivities of individuals and families to create "modifications in the variables of the environment." Ex-herders are taken to *"respon[d] to* them in a non-random way, in a systematic way" (Foucault, 2010: 269, italics added).

That is, local cadres increasingly pose the economy not simply as a question of development, but as an experimental medium through which human behavior might be provisionally choreographed.[19] The subtle manipulation of economies for roots, beyond all legibility to free market economics, could compel ex-herder participation, as if by the unlocking of an attractive force in the proper configuration of elements and environmental forces. In Alxa, stoking an inelastic demand for medicinal roots aimed to create a stable and guaranteed final buyer for roots, taking the volatility of markets out of the picture. These roots animate the forestry bureau's attentions and open its purse strings: their roots are a fulcrum around which money and state support move. For forestry goals, local officials have in effect rezoned Alxa as a tiny and highly controlled special economic zone (Ong 2006) specifically geared to attract and capture enterprises that would anchor demand for *rou congrong*, even before a critical mass of producers in the region existed.

The promise of a waiting buyer guaranteed by political manipulation of market conditions indeed sets up the economic environment for *rou congrong* cultivation as a supply chain with buyer but no sellers. Becoming a producer of the root was to occupy a structural position in a commodity chain presumed by buyers waiting for an as yet unproduced product. The economic environment created through state manipulation thus lures with its politically sustained incompletion, drawing ex-herders into the cultivation of shrubs and roots to occupy a niche in a supply chain ecology constituted across entangled but mutually irreducible motivations, forces, and economic and ecological forms (Tsing 2009, 2015) that can always accommodate more producers, and therefore more acreage of carefully maintained windbreak. This is a supply chain ecology that is complete but for the suppliers.

The market economy is itself often narrated and negotiated through figures of risk, demanding forms of embodiment and subjectivity immanent in the volatility of markets themselves (Zaloom 2004).[20] In Alxa, cadres work to expunge risk at every point through state intervention, especially through the artificial creation of an open and inelastic demand. Growers like Li worry less about market fluctuations in price than about the political will to maintain open and stable demand. Risk resides, then, not in the market *per se*. It is rather the effect of the careful attention that growers pay to the affects of the cadres that they interact with, or their pronouncements on sand control, forestry, and local economic development, which they listen to as coded messages on the longevity of state-sustained *rou congrong* demand. Trained on decades of splashy campaigns that come and go, they are "savvy about how to interpret politics and government," an ability "honed under socialism" (Rofel 2007, 22). The volatility of the market matters less than the mercurial whims of local heads of state, who themselves translate the coded languages of their own higher-ups into the lives of herders, growers, and herders-turned-growers. Their entrepreneurial action thus unfolds in a slightly different "environment," one that overlaps substantially with state attempts to create a bulletproof market, but is nonetheless careful in scanning the political climate for the first signs of a gathering storm in the state-buoyed markets for rou congrong.

In the meantime, growers like Li anticipate that as long as Beijing has dust storms, support for afforestation programs and the ecologies of investment (D'Avella 2014) around them will continue on in Alxa. And the state-built economy, more and more improbable in its idiosyncratic features, rolls on. It draws more and more ex-herders into its waiting niches, opening in the divergent entanglements of national environmental goals, local afforestation bureaucracies, and the holoparasitic timelines of roots that require one another to grow. An economy with guaranteed buyers means that the phenomena of over-supply, fluctuations in supply and demand, and even economic competition are filtered out. A supply chain complete with guaranteed buyers has the luxury of being voluminous, as infinite as the meteorological insecurity that girds state support, and as deep as its coffers. Where more production does not mean falling prices, there can never be enough producers. And the more *rou congrong* that ex-herders can be made to grow, the more *suosuo* will rise across the plateau, pressing its roots into sand and bristling its branches into the wind.

Ecological Stimulus Package

The creation of a highly artificial economic environment, from starting capital to waiting buyer, is posed by local officials as a means of orchestrating ex-herder behaviors. These interventions aim to leverage the capacities of markets and other economic phenomena as tools for directing economic behaviors that would double as vehicles of desired ecological construction. They pass through the conscription, through manipulation of the economic milieu, of root-focused entrepreneurs as accidental foresters, diligently attending to the growth of windbreaks as a groundwork for cash crop planting. And if foresters who grow trees for timber eventually kill their crop, the reliance on holoparasitic *rou congrong*, which requires the survival of its host *suosuo*, means that "forestry" here is about keeping trees alive rather than cutting them down.

To forestry officials, ex-herder behaviors are a crucial hinge between techniques of state-led economic development, long implemented to increase rural GDPs, and the accomplishment of centrally mandated afforestation quotas. These programs largely do not figure entrepreneurial, root-growing activity by ex-herders as proactive *action* dictated by liberated economic agents surfing the rationality of free markets. Rather, they center around the rationality of resolutely *un*free markets, which are closely managed by local officials as the immediate economic environment for eliciting changes in ex-herder behavior, and they are provisionally liberated from pressures seen as synonymous with capitalist markets as such: fluctuations in supply and demand, entrepreneurial risk, and even the profit motive. It opts instead for an environmental behaviorism that emphasizes "*reactions* to stimuli and not *responses* to questions" (Derrida 2008, 122), let alone liberal agency or the notions of environmental awareness that circulate in capitalist environmentalisms.

In this section, I explore the positioning of ex-herders as an idiosyncratic kind of environmental subject. The motivations and behaviors of this subject are not figured as self-originating choices of self-motivated rational agents in a free market economy. Ex-herders like Li and company workers like Luo alike describe their shrub-planting operations as a *response* to propitious economic calculations on a suddenly ex-pastoral plateau. However, they diverge on the character of these responses. The company foreman Luo, whom we met in this chapter's opening, sees a braiding of entrepreneurial opportunity and moral redemption vis-à-vis Alxa's new economy and coming ecology, respectively.

Li, in contrast, at best poses his own work converting pastures into windbreak *rou congrong* farms, as a matter of survival in the shifting winds of states and their markets. "There's nothing that can be done" (*mei banfa*), he shrugs. If he seeks to stay on his land, instead of following his neighbors into resettlement housing off-pasture, his turn to root-growing is overdetermined by a suddenly reconfigured economic environment. This ironically echoes state forestry officials' description of overgrazing as also an overdetermined effect of a prior economic environment.

In contrast, local officials see their economic behavior as a quasi-automatic *reaction* to a supplely controlled set of state economic controls, through which different outcomes can be choreographed. An environmental technique presumes the disarticulation of an action and its author. Cadres approach the proactive survivalist *cum* entrepreneurial impulse that they presume to ground ex-herder behaviors not as a matter of resistance to power, but as the very site where state intervention might be most effective. Notice how ex-herder action is understood not as the release of a proactive entrepreneurial impulse to take advantage of changing policy conditions, but rather as a specific kind of relational effect. The action of ex-herders, in this arrangement, is understood as a response, the desired outcome of the experimental modulation of an environment that can elicit it, as a behavior follows a stimulus. The environment, that is, is the author of ex-herder action, and ex-herder agency appears as what herders *cannot* not do.

The subtle distinction between response and reaction is crucial in understanding the specificity of this environmental subjectivity. Jacques Derrida notes that this distinction is crucial in founding the difference between animals and human beings, in general. He locates this distinction, "fixed within Cartesian fixity"—the image of the animal-machine—and in "the presupposition of a code that permits only *reactions* to stimuli and not *responses* to questions" (2008, 122 emphasis in original).[21] Behavior is not the expression of a capacity of individuals to act. It is instead framed as the after-the-fact effect of environmental conditions that are being manipulated in real time, balanced against the needs of plants and the reactivities of ex-herders. If an environmental subject acts, it is as the response to an authorship that inheres in the impersonality of its environment and the efficacies of the arrangement of its variables. Ex-herder action, then, is continually figured as a reaction that confirms that efficacy of the state in devising a set of environmental and economic conditions conducive to the specific vulnerability of their psychology.

Reaction, in Derrida's critique, reduces the complexity of behavior to "prewired behavior" (2008, 123) triggered by a stimulus. It is "what happens when one presses a computer key" (2008, 125). Response, in contrast, indicates an access to language, deliberation, and the possibility of a difference. This introduces governance as an environmental neo-behaviorism, where ex-herder agency appears only as a protracted range of possible reactions to a stimulus. They operate on a conception of responsiveness to environmental changes that cannot account for what Donna Haraway calls "response-ability," a "capacity for response" in a field of possibilities that is "not fixed by teleology or function" (2012) and thus remains open to the possibility of responses other than those anticipated by officials.

Mrs. Li, Li's wife, tells a story when I muse on how difficult it is to get a clear answer from herders and ex-herders about the size and composition of their herds. Smiling, she discloses an open secret about ex-pastures in Alxa. "It is not difficult to move a herd of goats behind hills and dunes when you hear the forestry police are coming for an inspection." The landscape and her ability to traverse it with the family's illegal herd, fragments the panoptic impulse of forestry department surveillance. Dunes, ironically, provide cover against the antidesertification police. Mrs. Li also notes that it is not difficult to pay off the forestry police, who inspect herds for compliance, with a goat or a sheep, which have become locally scarce and valuable *because* of the bans. She estimates that one goat per inspection is enough to keep the police at bay.

As her family diligently converts part of their ex-pasture to forest and does the groundwork that will contribute the coming harvest of *rou congrong* as a mechanism for meeting the forestry bureau's *suosuo* quotas, it also continues to keep a herd. They subvert the bans while collecting their attached subsidies. Instead of reacting mechanically to stimuli issuing from a shapeshifting state in the wake of a pasture-breaking shock, they have responded to new conditions by learning to manipulate it to other ends. Her insouciance and half-concern for the desires of local cadres is also a Spirit of the Camel. She keeps a troupe of small camels that roam far and wide, grazing wherever their tenacious appetites take them before they loop back to the *suosuo* forest, rising one season and planting site at a time against the spring wind.

It is often said that China has had a continuous history of five thousand years.[1] At times, the cliché expresses a pervasive cultural chauvinism through the profundity of civilizational time. At others, it is a way of neutralizing the tumultuous recent past into the gravity of deep historical time. Those involved in antidesertification work often raise the phrase as a subtle reminder that it is possible to describe environmental and human time in the same scale and time frame. "Five thousand years" admits no necessary distinction or sense of a mismatch in speeds and scales, no singular event of the collision of nature and culture.

I heard this phrase very often when I was with scientists driving through Naiman County in northeastern Inner Mongolia and its alternating patchwork of pastures, newly opened farmlands, semidegraded fields, and rolling dunes. It was deployed especially when I expressed surprise over how calm they seemed over the emergency of desertification. "Five thousand years" offers a trope through which they could explain to an outsider the ordinariness of this apparent disaster, subsuming its exceptionality into an account of ordinary environmental processes.[2]

Tingting, a young hydrologist at Naiman station, for instance, thinks of the depth of time not in relation to permanence but rather as a reminder of the very transience of the present as a moment in a longer cycling. As our van speeds past the new furrows of a freshly plowed piece of farmland in a grassland,she chides me for imagining that this opening is a violence against a pristine nature:

> All over China, places like this . . . have been settled for thousands of years, and the land is put to many uses and undergoes many changes. It is not like in America or elsewhere where land is plentiful—here, there have always been people and they have always worked the land in different ways. When we see these farmlands being opened up, this is something that has happened many times in the past. People farmed the land, and that changed the land and it deteriorates and underwent desertification. When desertification has happened in the past, grass has come back, and then people will come in and graze [their animals] here, and then, when the land is ready, others will come in and farm again. This has happened many times over thousands of years here.

She is not elaborating a specific catalog of events but, rather, the environmental form through which a temporality and process of landscape change can be imagined. The various kinds of landscape—dunes, pastures, fallows—rearrange as points in a process that extends backward and forward in time. Her account of the Earth as a long cycle of the rising and falling social, economic, and ecological regimes recalls both the models of ecological disturbance and succession through which the scientists see a dynamic Earth as well as the long cyclical returns of Chinese dynastic time (Levenson 1968). Geology and history are both matters of *longue durée*.

The opening of "new" farmland, a primary driver of desertification, for her, is a reopening, an indelible moment in a cycle that has lasted since time immemorial. It is proof of the perpetuity of the planet, even in this local quarter of itself, as itself a process of change. Such change is not driven either solely by human agency or by the workings of an extrahuman nature. And the concept does not require determining whether in a given event the one overtook the other, though Tingting and other researchers at the field station surely expressed concern over current changes, like the pollution of groundwater and the falling water table by drought and agricultural pump wells, that might tip the cycle off balance. Her telescoping of the variegated landscape into a tableau of a millennia-long, social-economic-ecological cycle makes apparent endings the necessary grounds for new beginnings. To her, the "disturbance" is already implied as part of the ongoing operation of the cycle, just as is its eventual rehabilitation. It is the normal disaster.

At the center of her way of seeing the Earth, where succession has shifted from an ecological principle to a framework for time more generally, is a relentless, cycling change that binds together the endurance of China and the endurance of the Earth. Tingting's earth is not an object but a process, gathering and bearing the effects of economic, ecological, and geophysical processes. "China" and its difficult land urge a thinking with perpetual cycles that repeat with all the spontaneity and mystery of the changing seasons or the rise and fall of governments. In this long China, the earth is the substrate of its own time. Tingting offers an account of a landscape that has long been embroiled in human practices and yet cannot be called anthropogenic if that term indicates human life as an excess over environments rather than a moment in them. In her earthly process, there is no need to understand the environmental present or future as a rupture in time nor to ask for five thousand years to bear itself as a history of trespasses.[3]

Apparent end points become the proof of a cycle beginning again, and civilizational chauvinism stages both the congruency and mismatch of political and geological deep times. The touch of their time lines displaces and extends them both, and Tingting proposes a cycling of political, material, and environmental changes, repeating over and over, from time memorial perhaps into the future. All this has happened before, she seems to reassure herself to an outsider, perhaps proposing these five thousand years to cushion the hushed conversations about contaminated groundwater, the falling prices of agricultural commodities that speed the opening of new farmland, a general sense of the present as social and environmental hiccup becoming a seizure in the cycle of social-environmental succession.

3 Holding Patterns

EARTHLY AND POLITICAL TIME
AT CHINA'S DESERT SHORES

Ecologist Li Ming and forestry official Tian are standing with me in the footprint of Qingtu Lake.[1] Once the anchor of a verdant oasis of reedy wetlands in Gansu Province's Minqin County, the lake today is an expanse of sand that extends past the horizon. As late as the 1950s, Qingtu Lake sprawled luxuriantly over four hundred square kilometers, sixty meters deep at its lowest point. In the span of decades, with groundwater drained in the utopian social-agricultural experiments of high Maoist socialism, the lake disappeared completely, leaving a carpet of mobile, alkaline sand. It swirls around our heels like water lapping on a still-remembered shore. It pools around our boots. On the spring day that Li, Tian, and I have come, we walk across the lake bed, wading shoulder deep in absent water.

Minqin County is dotted with places like the former lake, where the relentless pace of desertification has disjoined toponyms from the landscapes they named just decades before.[2] Minqin, once an oasis en route to the eastern terminus of the Silk Road, is wedged at the sandy nexus of three modern interior provinces. Their borders cut an unlikely jigsaw into desert: Gansu, Inner Mongolia, and Ningxia. The county is today an archipelago of places scattered like islands in what many of the state foresters that I speak to call the *shahai*, the sea of sand. The oceanic turn of phrase, *sand-sea,* is a metaphor that evinces possible meanings of sand and desert in Minqin.[3] It draws attention not only to the sheer size of new and active desert lands, but also to their oceanic activity. Loosed on the wind, desert sands gather into dunes that flow like a slow liquid, land swallowing land.

In this sand-sea, places like Minqin exist in the grim, anticipatory state of the not-yet-buried. As two deserts threaten to run aground over the oasis, scientists scramble to devise methods for modeling dune drift in the region (Sun et al. 2005), projecting timelines for Minqin's ultimate burial in the coming decades. For China's State Forestry Administration, the disappeared Qingtu Lake has become a powerful cautionary site in the central government's fight against desertification. Sand-control campaigns across China's desertification hot zones have become anxious stagings in the state's combat against mobile sands, which now afflict a quarter of China's landmass and periodically rise into its skies as dust storms.

Today, exposed lakebed sands make the air a translucent haze that stings our eyes and scratches our throats. For the forestry official Tian, the lake is an image of Minqin itself, its disappearance the county's burial foretold. From our vista point on the lake bed, Tian, who is in his fifties, remembers the waterfowl that have long stopped coming. He conjures disappeared waters in memories of taking his children swimming here. "The lake contracted, slowly and then quickly," he remembers. "Where it dried, the sand on its bed was free to move everywhere" on the lightest breeze. "With no water the farms went dry. Even if they did not become sand, it was impossible to keep sand from burying the fields." In his telling, sand has become a mineral architecture for remembering. It "holds geological memories in its elemental structure, and calls forth referential memories" (Agard-Jones 2012, 326). In its texture, color, and dryness on the skin and in the throat, in its gritty lodging in a fold of fabric or flesh, exposed sand structures a catalog of environmental changes and disappearances.

Sand, for Tian, renders time into recursions. Remembering in this way is also a foretelling. Materializing across fluid and solid phases, it moves, accretes, holds momentarily steady, or tends to dissolution. It draws together past and future burials, each as "an exception, shall we say, that announces itself only as an example" (Song 2012, 131). The sands on the lakebed recall and anticipate other disappearances; past burial events appear as an earthly momentum, and Tian recounts sandflow to offer a cautionary materiality. "In ten years, twenty years, or five years, the sands will swallow Minqin, too," Tian says, pushing his hands together to mimic the press of the two deserts that flank the sand-locked county. This pressing is a foretelling of Minqin's future from the perspective of a past ending.

Each burial confirms the capacity of sand to bury. Sand, for the forestry official, offers a medium through which pasts and presents come to form. Marking time in sand flow diffracts any unified present across desert China into a crystalline noncoevality of stages in a singular geophysical process playing out across disparate sites. Its past and future motions link memory and prognostication as the repetitive realization of a process already imminent in the geodynamics of the unrestrainable desert. As lake-bed sand slips through my fingers, it disperses with the breeze into a mineral fog. Li Ming, my ecologist friend, explains, "In places like this, we are no longer confronted with the problem of development [*fazhan*], but rather the question of existence [*cunzai*]."

———

In this chapter, I follow Tian, Li Ming, and their colleagues across China's state sand-control apparatus in order to learn to tell time through sand. Together, we take up sand as what Stefan Helmreich (2011, 132) calls a "theory machine" that operates at the interface of material processes, environmental change, and the chronological conditions and tactics of political intervention. Telling time through sand's properties is, thus, a way of giving form for "near futures" (Guyer 2007, 410), poised among, and often veering away from the immediacy of omnipresent crisis, insurgent disaster (Massumi 2009), and modern eschatologies of environmental collapse, anchored by proclamations of the end of the world itself.

I ask how sand-control programs in practice make governmental control of and intervention into time a site of chronopolitical experiment, one that takes shape in the multiple timelines and temporalities that can be recovered out of sand's material excess. How might the politechnics (Anand 2017) of sand control also generate an archive of speculative futures that undermine the disparate "chrononormativities" (Freeman 2010) that govern *both* the state-sponsored progress narrative of Reform and Opening and the end-times-thinking of climate pessimism? The soaring and collapsing futures of the state and the Anthropocene, like a stepwise gear, are reshaped through sand as a material that variously accretes, buries, oscillates between motion and stability, or provides a habitat for geoengineering plants.

The multiple temporalities that emerge in sand-holding work also rework the chronopolitical stakes of environmental governance. By *chronopolitics*, I mean the various ways in which the political does not merely operate in "empty, homogenous time" (Benjamin 1967, 261) but, rather, tacitly and sometimes explicitly makes the manipulation, acceleration, or projection of time both the condition and ongoing goal of political and governmental intervention. I approach encounters of sand and state at Minqin's desert shore as seams of chronopolitical experiment, wherein earthly temporalities and the horizons of statist time are folded into one another, yielding an array of unexpected chronopolitical forms.

The difficulty of stabilizing sand's tendency to motion with wind, for instance, upsets any straightforward account of environmental modernization, with its teleologies of eventual control and state triumph over its unruly territory. Instead, sand engineers coproduce time through their grappling with the properties and agencies of sandy landscapes. Straight lines arc into loops and inversions, or at other times, loops become remarkably teleological in their redeployment as counter-temporalities in a terrain of protracted sandy intervention. As sand patterns into the projects that aim to hold it, other figures of state action in time torque out of the linear progressions of the state calendar: waiting, repetition, exhaustion, erosion. The chronopolitical stakes of state intervention are reset, realigned, and remade in encounters of sand and state, where dynamic landscapes and experimental state practices meet in a clash and emergence of chronopolitical patterns.

I am interested in how attunements to geological time in the ecology of institutions that center sand control refract geological and chronopolitical temporalities into one another. This mutual refraction of political and environmental temporalities creates a series of what I approach as holding patterns—reconfigurations of landscape and temporality that disorient the promissory futures of the state into a series of stranger shapes in time. Such patterns manifest as environmental engineering projects that both take up and intervene in sand as a material in time and space. In other ways experimental techniques and new kinds of attentions reshape their temporal forms and horizons around the propensities of sand as a mercurial substance.

METAGEOLOGIES

The question of the relation between political and earthly temporalities has mattered powerfully in ongoing attempts to capture a sense of a changed planet and its futures. This problem, which projects new Earth futures through the emergent geomechanics of our damaged planet, has involved, in many cases, two scalings-up: both to thinking of the planet *per se* as the relevant arena for considering environmental questions, and, importantly, for redoubling on an abstract vision of humanity. Following interlocutors doing their best to rout, hold, and reshape sand's futures, we attend to the prospects of a chronopolitics patterned with the planetary dynamisms of this Earth.

Dipesh Chakrabarty (2009) approaches the Anthropocene question and a suite of contemporary environmental processes more broadly, as they open a problem space over how to give form to futures in the obliteration of a cleavage between human and environmental history. If "it is being claimed that humans are a force of nature in a geological sense" (2009, 207), this breach "in the binary of natural/human history" (2009, 205) raises profound historiographic questions. This includes, shockingly, what may be deemed a kind of secular eschatology that explicitly charts a new kind of history that poses environmental futurity as an engine of civilizational and other collapses. In an increasing alignment of environmental processes with doomsday, environment is increasingly the name of a powerful process that vibrates with the sense of an ending.[4]

If modernist modes of history making depended on both the neutralization of natural and divine processes into the theoretical infinity of the open-ended calendar (Koselleck 2004, Luhmann 1998), here we witness attempts to grapple with the uncanny sense of a reconvergence of geological and historical time.[5] The Anthropocene thesis can in this sense be understood as a way of asking how environmental and political temporalities meet. In any case, earthly processes and their more-than-human temporalities appear beyond questions of the description or modeling of landscape processes. They are increasingly reframed as a way of planetary fortune-telling whose very gravity may very well warp and torque the political temporalities that have long figured nature as a problem to be solved or a chaos to be pacified.[6] If, today, the status of humanity as a geological agent is at stake and "geological time" oscillates wildly between the depths of deep time and the accelerations of the Anthropocene as a geostratigraphic age, it is incumbent upon us to explore how the encounter of environmental and political timescales are already producing unexpected temporal repertoires.

For the scientists, bureaucrats, and engineers in the Chinese state's dispersed apparatus of sand control, to attend to sand is less to tell a history, a specific catalog of events, than it is to frame a metahistory. "Metahistory" is a term I borrow from Hayden White (2000), who proposes it as a schema to investigate how historiographic writing depends on formal and narrative conventions that generate senses of momentum and expectation. Metahistoric genres and forms generate narrative expectations and futures in a story. These forms are out of time, but they structure time and create temporal expectation, including in well-known conventions of ethnographic writing as an anthropological literary genre (Fabian 2014). As David Scott argues, specific generic forms hold any particular political present in relation to "the salience of the horizon in relation to which it is constructed" (2004, 19). For Scott, these are more than questions of aesthetics. They are an injunction also to recircuit the political as such by attending to modes of emplotment that have been unwittingly inherited into ethnographic thinking and then experimenting with metahistorical formats that yield other stories.[7] The experience of time itself thus cannot be separated from the array of forms through which temporality can be shaped. The spaces of relation between different genres of time-making emerge as a terrain of chronopolitical experiments.[8]

Tracing these patterns ethnographically as they emerge and shift in sand-control works and infrastructures allows us to explore how environmental and political time might come into an array of concrete relations. I therefore turn to sand control as a site for experimental entanglings of earthly and political time in which "environment" stands neither simply as a pure material disruption of megalomaniacal state designs on mastering "nature" nor coaxes an elegy out of sand that would, in any case, replicate a contemporary tendency to see only endings, disorder, and collapse in the very texture of a changed earth and sky.

In China's desert interior, haunted by tales of buried kingdoms and their whispering ghosts, the desert disappears history, as if burial were a matter of creating perfectly preserved archaeological archives: ruin as geological technology for stopped time.[9] But here, desertification and attempts at its control offer a method for making time emerge in the ruckus of different accounts of how sand reshapes political temporality between different ways of knowing, encountering, and intervening into sandflow. And if Anthropocene and climate-change futurologies have tended to register a *unification* of environmental processes in pointing to future (or already-here) havoc, here, "geological time" is an open question of how sands, ecologies, and a state reckoning its own prospects against advancing dunes collaborate to make other earthly timelines viable.

High-flying Chinese state claims that ecological construction programs like sand-control infrastructures manifest a "teleological graduation into eco-rational modernity" (Yeh 2009, 884) give way into a complex repertoire of ways of accounting, engineering, and manipulating the play of multiple time lines discerned in the physical properties of sand. They not only return deserts into the flow of history, but also reconfigure the flows and arrests of sands into a substrate for other ways of shaping futures. Mobile dunes, blowing sands, and desertification are sites where we can trace emergent alignments of politics to the inorganic afterlives of the broken land. How, in practice, do specific environmental processes and materials provide a repertoire for chronopolitical experiments? If sand encroaches into space, how, too, can it encroach on time?[10]

CHRONOTOPES OF SAND AND STATE

> Sand not only flows, but this very flow is the sand.
> Kōbō Abe, *The Woman in the Dunes* (1991, 99)

In combating sand, state scientists and engineers organize programs of state landscape engineering whose temporal horizons and chronopolitical techniques rework the futurism of the Chinese Communist Party's political imaginary. For the state theorists, the ecological problem suggests passage into the next historical, "ecological" stage of Chinese socialism (Li & Shapiro 2020, 6).[11] Desertification politics, however, take shape as a site for the proliferation of environmental-material temporalities that undercut any sense of statist politics unfurling in a singular form. Its materiality conflicts with the "changing materiality" of economic development and infinite improvement in living conditions, only phantasmatically captured in statistics (Liu 2009, 38), that have been at the center of statist futures in Reform Era China. Geophysical and ecological temporalities traced through sand complicate the soaring statist futures that have long been a mainstay in Chinese politics. The future orientation of modern Chinese politics and, more broadly, a contemporary investment in speculative, anticipatory things (Adams, Murphy, and Clarke 2009) have long made the future a tactical achievement of governing.

As Ann Anagnost has argued, modern Chinese politics must be tracked through "tactical plays on time" (1997, 7), where state practice did not merely have temporal dimensions but was indeed explicitly conceived and practiced as action on time, in the distinct register of speeding up the arrival of the future. In the People's Republic of China, there has been a succession of official futures, from Mao's worker's paradise to the new promises of consumer paradise (L. Zhang 2010) in the vertiginous promises of infinite economic development.[12] However, such chronopolitical gambits have also, at different moments, aligned Chinese historical experience with a sense of continually frustrated anticipation. Lisa Rofel writes that experiences of Chinese modernity take a structure of frustrated anticipation in the "repeatedly deferred enactment marked by discrepant desires that continually replace one another" (1999, 9–10). In Reform China, ethnographers chart a proliferation of political and personal futures in the new possibilities of consumer and desirous futures in the Chinese amalgam of authoritarian control and market experiment (Ong and Zhang 2008, Rofel 2007).

For state bureaucrats, ecologists, and sand-control engineers, the earth appears as a mobile quantity, the material substrate of many futures in a tangle of engineering techniques, ecological interventions, and anxious hopes for topographical control and the revivification of a sand-choked future. As sand flows in dry waves or sustains successive plant seres, it appears rather as a material in and at the cusps of multiple processes and at the center of political and technoscientific experiments. Noting the multiplication of earthly temporal forms is a matter of more than emphasizing the ontological multiplicity of worlds. Rather, we might follow Annemarie Mol to insist that "attending to the multiplicity of reality is also an *act*" (2002, 6, emphasis in original). And, embedded in and emerging through state designs to control of sand, the enactment of sand's multiplicity is an exercise in building a chronopolitical armature so that some sand time lines can be twisted into others. Interdisciplinary research and practice on mobile deserts work to foment temporal multiplicity in order to strategically open sand as a chronopolitical terrain, opening spaces of political and environmental futurity amid the foretold burials of cities and lakes by sand. To think with Reinhart Koselleck, "What follows will therefore seek to speak, not of one historical time, but rather of many forms of time superimposed one upon the other" (2004, 2), with a particular attention to sand-time, as it patterns into the temporal grounds and horizons of state action.

WAITING FOR THE HORIZON

From a pavilion atop the reservoir that feeds Minqin County, the county's main town is visible as an island between two expanding deserts. Minqin is acutely vulnerable to engulfment by sand pressing against the main town on both sides (fig. 7). The central declaration of the county as a key zone in the nation's fight against deserts has drawn expertise, resources in the billions of *renminbi*, and political attention to this unlikely place, already 95 percent covered in sand. This vista is the first that visitors to Minqin get as they emerge from the sheltering colonnade of windbreaks protecting the only road into the county. In the disappearance of local surface water, this reservoir holds and drains the diverted Shiyang River. Its waters feed the antisand forestry programs that have sprung up around the town like medieval fortifications, a moat between city and sand.

As sand threatens the very possibility of ongoing habitation, Minqin has become an exemplary landscape. Minqin's predicament conjures—by proxy and through reference to prior burials, like its dried lake—a

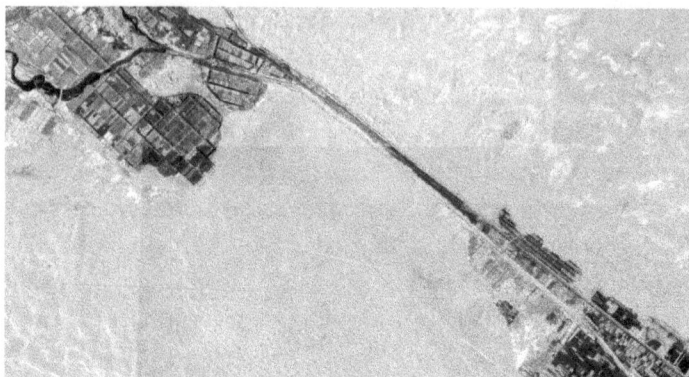

Figure 7. Minqin County main road. Google Earth. Prepared by Micah Hilt.

pedagogical vision of China splintered into various moments in a timeline of coming burials. Minqin matters as a geophysical haunting. The officials that gather here describe its landscape in reference to the mobile deserts that loom just one hundred kilometers outside of Beijing's city limits. If Minqin takes its place in a catalog of other disappeared places, it also stands as an anticipatory echo of a possible disappeared Beijing.

I have come to Minqin this time with a group of forestry officials making their pilgrimage to what has become a Mecca in the world of Chinese antidesertification politics. They are a group of technical, scientific, and bureaucratic officials who have been charged with holding the desert at bay. Some, like Tian, are local forestry cadres who are invested in saving their hometowns. Others, like Li Ming, my ecologist friend, have come from other places in the institutional geography of state sand control, including officials from Beijing.

In central-government parlance, the movement of the sands is a "swallowing" (*tunshi*).[13] Minqin sits at the throat of the two deserts closing toward it from three sides. It is a place that can only be said to still exist. Everywhere, the city's predicament is posed in relation to the fearful future moment when the two deserts touch, rendering the city uninhabitable. The looming future ending drives changes in the social and physical landscape of the region. It has spurred a general reorganization of government in this ex-oasis through the politics of sand control.

Figure 8. "Minqin must not be allowed to become a second Lop
Nur." —Wen Jiabao. Minqin County, Gansu Province, November 2011.

The massive investments in controlling the desert are conditioned
by and elaborate this sense of an ending that is always, literally, on the
horizon. Minqin's slogan, visible for teams of visiting cadres and forestry
workers, for instance, narrates political investment and urgency in reference to a future that must be averted at all costs. Penned in the calli-
graphic hand of the geologist-turned-premier of the CCP, Wen Jiabao, the
slogan demands that "Minqin must not be allowed to become a second
Lop Nur" (*Jue buneng rang Minqin chengwei di'er ge Luobubo*) (fig. 8).
Like the lakes that once sprawled over Minqin's oasis, the first Lop Nur
was for centuries a massive inland sea in the neighboring Xinjiang region.
In 1973, it was declared officially dry, leaving a massive alkaline footprint,
which has become a major dust-storm source in western China.

Wen's Minqin is already located as a point on the time line of its own
disappearance, one that approaches the city in the backward shadow of its
coming negation. In a speech upon visiting the choking oasis, he urged,

"We must win 'the fight for Minqin,' and by no means should we let it vanish from the map" (*Xinhua* 2007). The slogan enacts a call to a politics that approaches the future as an event to be held at bay, that which must not be allowed to happen again. It does so by linking Minqin's future to a past disappearance and both to the development of deserts in general. A politics against the desert takes, in Minqin, the structure of a negative command. The triumphalism in state narratives of infinite progress is reshaped through the ongoing work of holding sand against its powerful tendency to spread. Such a politics in sand no longer addresses itself to a utopian future but merely to the aversion of an ending already in process. Nor is there a sense that the state can do anything better than hold the desert at bay in a tentative stalemate that may at any moment tear open. Wen aligns political intervention with preventing an event that has already happened and, left to itself, is on the brink of happening again.

In the slogan, political action addresses a future that hurtles closer by its own progression. It figures a China where sand-bound places arrange as a movable history of disappearances, each an iteration of a general desertification timeline. In this sense, the combat of sand and state that has defined Minqin for forestry officials must be understood as a rehearsal for other places, especially Beijing. If cities in China strive to repackage themselves as models of sustainable urbanism (Hoffman 2011), Minqin County takes its place in a long history of political example, a state-sanctioned pedagogical genre in which designated citizens, cities, and revolutionaries are raised to the status of models to be emulated. In the slogan, Minqin, in a subtle turnaround, stands as a negative exemplar, a model of what must not happen. The cautionary character of burials, each a demonstration of the power of sandy lands to bury, is thus not simply a question of the vanishing of a single place from the national map but the possibility that sand might disappear other places.

The group of visiting officials climbs stairs like pilgrims to special state-built viewing pavilions, which recall Chinese garden architecture, with its carefully framed vistas.[14] These views orient a view on the landscape through a visual architecture of picturesque environmental threat. The sheer enormity of a desert appears from these pavilions as a city-killing force, crashing against and over the sandbreaking engineering projects installed and continuously rebuilt under the auspices of Minqin's county forestry bureau. As a series of state-sanctioned vistas, these pavilions, each emblazoned with the premier's warning slogan on a stele—*Minqin Must Not Become Another Lop Nur*—emphasize endangerment by sand rather than the technological sublime (Nye 1994) in state mega-engineering projects. They forebode. We leave the promontory at the reservoir's edge and embark on an inspection tour that will take us past half-buried barricades at sand's edge and into the desert, where sandbreaking grids built of straw, nylon sandbags, or green plastic netting designed to function and look like sand-holding vegetation undulate with the ebb and flow of the dunescape.

For the foresters, the landscape is a monument to its own movements and changes in time. Different places become exemplars of stages in the life history of desertification. The series of sites string together as moments spread in a tableau over the county's sand drifts. Their progress can be read through the evacuated villages and buried fields that extend far past sand's edge. The absence of people, due to the depopulation campaigns that have been an emblematic feature of local anti-sand programs, appears as proof of the power of the desert. The rutted-out trails that balloon off paved roads, where sand has encroached in smooth pools, catch the shape of the wind. Buried and toppled sand-control infrastructures mark the power of the desert to continue.

To these forestry officials, the land is movement; it is a speed. Speed here is not quickness but, rather, a relation between time and distance. It does not stop; it only rests. The land's motion and stability do not appear as opposites but as two possible expressions of this time-space of sand, a substance that is a flow. Sand, rendered as material velocity, incites a reflection on speed in counterpoint to the "contraction in time, the disappearance of territorial space" (2006, 156) that Paul Virilio identifies with modern technologies. At the reservoir's ridge, engineer Liao speaks to the group, framed by the stunning backdrop of a town pressed on both sides by walls of sand. "The sand moves an average of ten meters per year from each side, sometimes reaching twenty meters," he estimates. He presses his hands together to simulate the two deserts. "When the two deserts hold hands," he says, "not only will Minqin disappear, but this whole region will have changed, in just decades, from a lush oasis to an unbroken plain of sand." Where geological epochs arrange vertically in stratigraphic layers, Liao marks and projects the passage of time moving horizontally as an advancing contour on the earth's surface. This horizontal spread, from Minqin County's main town, appears as the approach of a dry ocean.

Forestry and sand-holding programs appear as techniques for intervening in this rate by modulating sand's speed. Liao and the foresters, invited to imagine Minqin's future as a process of inexorable burial, articulate the desert's speed as one that is both fast and slow. When terra firma can shift into a mobile substance, any movement is too fast; and yet, because Minqin's burial may yet take decades, the problem of desertification is one that remains forever in a future that is at once both too close and too far.

Rob Nixon suggests that environmental processes like desertification fail to become properly political because such "slow violences" fail to breach the threshold of attention for a politics oriented toward spectacular events (2013). Speed, Nixon suggests, is a condition of politics as such, and more importantly, it is a temporal disjuncture between the slowness of environmental catastrophe and the short attention spans of contemporary politics. But such a process and politics elude the binary of fast politics and slow environment implied by "slow violence," with its straightforward mismatch of environmental and political speeds. Indeed, Minqin forestry engineers find the county caught in a period of waiting, wherein the end is always present and yet still not immediate. Joseph Masco explores the depiction of nuclear holocaust in early Cold War America to argue that the continual state announcement of the end of the world in Cold War cultural politics centered the "national contemplation of ruins" (2008, 361) as part of the political reorientation to disaster. Constant depiction of future ruins was a technique for the conditioning of an unsettled everyday affect, a sense of living emergency.

In Minqin, out of the many ways in which political practice and lived time are oriented toward a coming disaster that is both too fast and too slow, the present has become an intermediary period. While for Masco, the urgency of nuclear threat animates a sort of frenzy, one that persists in contemporary American antiterror campaigns, in Minqin, it is a different kind of urgency: waiting. This waiting must be prolonged. Preventing Minqin from becoming a second Lop Nur or a cautionary vision of a future Beijing means that sand-control efforts locate the present in the fearful anticipation of desert encroachment. They are looking for a holding pattern, forestalling the future by engineering the landscape into a stasis carved against the progression of sand.

The remaking of Minqin's social and physical landscape is therefore a means of engineering space to engineer time. Sand-control infrastructures that stall the flow of the land are part of a more general calculative logic that aims not to obliterate the desert as such but, for now, to stall it in a holding pattern actively poised against sand's futures. A slowed landscape equals a future deferred. Minqin has become a showcase for the antidesertification engineering techniques that, under the implementation of the forestry bureau, have sought to protect the city by a set of practices of *zhisha*, sand control, or, more colloquially, *yasha*, pressing sand. Around Minqin, there is an engineered landscape at and beyond the sand's edge, 330 kilometers of barriers positioned to cut sharp edges back into a dry liquid landscape. Farther in, there are rolling dunes engineered into tentative stability, part of a spectacular infrastructuralized landscape where the encounter between sand and state is staged.

STATES OF DISREPAIR

The ongoing labor of state-led reconstruction paradoxically means that, in forestalling the end of the world, there is no end in sight. At Laohukou, where the Tiger's Mouth mountain pass funnels air currents into sand-blasting jet streams, Liao, a veteran sand engineer, wends his truck off the road into the curves between dunes, where we see construction in progress. Trucks are piled high with bales of straw, collected after the corn harvest in other parts of the region. A corps of men, mostly ex-farmers who now work the dunes that have buried their fields, dig the straw stalks into the ground, tracing out an undulating grid of low windbreaks. Breaking the smooth dunes into rolling grids, I think of the seams of quilts holding patches in place.

Such physical barriers are replicated in a number of materials, from other kinds of farm waste to burlap or nylon fashioned into sandbags or low walls (fig. 9). "Biological methods" like tree planting or aerial seeding are prohibitive this far into the desert because of the low water content and general shiftiness of the soil substrate, and in emergency situations,

Figure 9. Nylon windbreaks for sand control. Photo by author.

chemical methods, mostly involving the spraying of petrochemicals directly onto sand, laminate loose sand into a self-adhesive sheet. They turn the dunes into a rolling barrier against themselves, and as time goes on, sand forms into slopes on the inside edges of the squares rather than moving across the land.

Forestry engineer Xu, who works for Minqin's county forestry bureau, coordinates the logistics of rounding up materials and labor for construction. He laments the ephemerality of all of the infrastructure that today proudly announces itself with a slogan written in cornstalks: "Fix Sand and Block Wind! Protect our Home" (figs. 10 and 11)! "Grass is organic," he explains, "and so it rots and it is subject to erosion. If it is an especially dry few years, the squares may last for four years, but they may disappear as quickly as two." Nylon is more durable, but more expensive. And even if the grass or nylon squares can outlast the process of their own decay, he explains, it is likely that they will be buried before that time comes.

Figure 10. Newly-laid slogan, "Block wind and hold sand (*fangfeng gusha*)" surrounded by two-year old grid. Minqin County, 2012. Photo by author.

Figure 11. "Welcome to [Illegible]" in eroded windbreaks. Minqin County, 2012. Photo by author.

Xu's job is also the coordination of new construction and planting, which moves one step ahead of the sand's leading edge to slow the flood. "We will be doing this for years," he confides, looking over the organic and abiotic sandbreaks that have made the landscape an unnatural infrastructure (Carse 2014). Construction implies the work of an ongoing repair and maintenance, and then rebuilding over the buried ruins of earlier projects. The work is labor intensive, and sand control has become a major driver of the economy, "catching" those no longer able to work their fields. Closer to the city, the quick-growing poplars and shrubs planted for sandbreaking demand continuous replanting and watering and ultimately have a notoriously low survival rate—as low as 30 percent nationally.

Holding the present in place is a continuous work, an interminable expenditure of materials and labor for an infrastructure that is only ever evident in its breaking down and its need for repair (Larkin 2008, 235–36; Star 1999). It takes more and more to make sure nothing happens.

For those who oversee and build straw grids, their smooth and already eroding lines are testaments to the vast amount of continuous human labor that must be enrolled and mobilized for infrastructures that are breaking down the moment they are installed. Straw grids and other windbreaks are more *re*-built than built. The clean lines of new construction signal that somewhere just below foot, another project has been buried. I learned, building grids myself, to see in new projects the "workings of disrepair" (Chu 2014). Local forestry bureaus do not promise a final deliverance from sand. Their work is structured by repetitive burial and building. It is not to escape cycles of burial, but to contend with them with cycles of work. They struggle to rebuild the quickly eroded infrastructures through which another few years might be eked out against the sand.

Sand displaces that grand state future into an ongoing now, with a politics that aims to continually maintain an unchanging present. It is, of course, the flows of sand through which coming time takes shape and not the promises of a state that cannot do much more than insist on flagging capacity to hold back midnight for another season. It patterns an environmental-political endurance that neither knows nor expects release. It no longer demands any transcendent future, but only the simple continuation of existence. This is a politics animated through the urgency of keeping time and space in a condition of still-waiting, fiercely remaining in a condition of suspension that must not admit a final transition. Still on the horizon, the future is the space of a holding pattern that, if possible, can persist a few more years.

SUCCESSIONS

It is after dinner, but the sun has barely begun to set in the big, high summer sky. Naiman's desertification research station in eastern Inner Mongolia, one of a dozen scattered around China's desertification hot zones, buzzes into life as people settle in for evening activities. Summer field research feels like summer camp for the teams of ecologists, GIS and remote sensing experts, hydrologists, dune scientists, and engineers.

Li Ming, my ecologist friend who will take me to Minqin months from now, and Little Tu, a first-year ecology PhD student, suggest that we walk the grounds behind the institute. Students and young professors like these conduct the bulk of research at the station. As dusk approaches, they want to climb the watchtower, where we can watch the sunset while eating apricots from the tree near their dormitory. The paths on the institute's grounds curve and meander through what Li Ming says were sand dunes just years earlier. Now they are dense and unruly with branching shrubs, stands of small trees, and low, weedy grasses that spill over the footpaths. Climbing the watchtower, Little Tu muses that one day, before summer ends, they could light fireworks off the watchtower's platform.

Looking back over the half-dozen main buildings of the field station and then turning to face the horizon, Little Tu notices that the vegetation changes in thickness and density. She discerns a botanical gradient, from trees to shrubs to grasses, blotching into patches on the sand before thinning into bare yellow dunes. Li Ming, the senior of the two desert ecologists, just starting his first position as a full-fledged state ecologist for the Chinese Academy of Sciences, listens to her. He left his hometown in China's subtropical south with a romantic longing for the open landscapes of northern China's steppes and, now, deserts. As she speaks, he interjects here and there to identify key grasses, shrubs, and trees, in Chinese and then in Latin: *shami, yanhao, shaliu; Agriophyllum squarrosum,*

Artemisia halodendron, Salix cheilophila, interpreting the patches of plants (Tsing et al. 2019) in the late summer dunes.[15]

Little Tu, who is doing field research in Naiman for the first time, has the unenviable task of processing sand samples through a sieve with the back of a spoon for later experimentation. It is not my first time wandering on the grounds with researchers at the station, for whom the changing communities of plants are a living example of the powers of desert ecology to root dunes into place. Little Tu knows enough to know that the species that Li Ming names occur naturally on dunes in the region, and she knows that some of these plants, like the dwarf sand willows that have rooted closest to the station's buildings, are strategic species for dune-stabilization projects. Plant growth is a stabilization technique. Little Tu wonders aloud, "How many different kinds of plants does it take to fix a dune?"

"The number of species is not the important factor," Li Ming replies, shifting subtly into a pedagogical mode and now addressing both of us. "From here, we can see several species, all at once, so it looks like a single community of plants. But look, the species are not evenly distributed— there are trees there, but not at the sand's edge." Tracing his finger across the land, from the open dunes through the gradient of plant species, he invites us to see the subtle progression in plant communities as various discrete moments, plotting out sequential years.

Li Ming exhorts us to see the various communities as slices of time, a simultaneous diachrony across the dunes, with its beginnings on the open drifts and its endings in the stands of sand willows. He continues, "When we come back next summer, the way that the plants are distributed will be different again—trees will replace some of the shrubs, and shrubs will replace some of the grasses. The *shami* grasses will be denser, and there will be new *shami* in the sand. The ecology is more unstable and more fragile where there are bushes instead of trees, grass instead of shrubs, and sand instead of grass. It is still developing, and where there are trees, it is more stable. We are not interested in the number of species, exactly, but in the stability of the ecosystem." The graduated composition of species on the dunes is a way of emplotting dunes into a process of vegetation change.

This process is a version of the classical formulation of ecological succession, first pioneered in Cowles's (1899) studies of vegetation on the dunes at the shores of Lake Michigan in the late 1800s and then the Clementsian paradigm (Clements 1916) with its organismic underpinnings.[16] Ecological succession charts a process in which dunes can be imaged as passing through successive botanical stages. According to the story of succession, on a denuded landscape, pioneer species establish and create the conditions for others. They are subsequently replaced by successor species that, in turn, create the conditions for their own overtaking and replacement by new species. Each stage stages the next, so that weedy pioneer communities are already inhabited by the promise of stable climax communities.

Ecological succession is a way of seeing sand at the intersection of multiple environmental processes, some of which can be mobilized against others; it is both a way of telling time on dunes and giving form for potential action into that time. Research in Naiman is an interdisciplinary enterprise, oriented toward developing dune-stabilization techniques. Its original incarnation, in the 1960s, was as an environmental-engineering institute, coordinating sand-control projects alongside desert-crossing Tongliao-Beijing railway, in a time when sand encroachment caused up to fifty derailments a year across the country (Wang and Zhao 2005). Principles of ecological succession allow state ecologists to see bare dunes as future forests and to pose the mobility of bare dunes as the starting point of an ecological process that will culminate in stabilization. At the desertification research station ecology is a science for understanding the dunes as shifting arrangements of biotic and abiotic things. It is at the same time a tactical element in the enterprise of dune stabilization; ecology toggles between scientific practice and engineering technique. This doubling of ecological time is a research objective.

For scientists at the station, the ecological stabilization that Li Ming describes is fungible with the stabilizations demanded by sand-control programs. Further into the process of succession and the approach to a stabilized ecosystem, plants tend to establish more densely, and the sand-securing capacities of deeper and more robust roots increase. Indeed, on the dune-surveying trips that groups of students and scientists make each day, documenting plant species and estimating vegetation cover on a square meter of surface sand are ways of locating a dune on the quadripartite scale of dune mobility. Whether a dune is mobile, semimobile, semifixed, or fixed can be determined by the character of the vegetation on its surface (fig. 12).

Figure 12. Li Ming estimates plant cover, 2011. Photo by author.

Ecological and geophysical stabilization appear as faces of the same sand-engineering process, and succession describes both an ecological process and a potential geoengineering technique. On moving dunes, then, ecological succession opens ways of mining the sand for its biotic potential to unearth futures that oppose the unending advance of dunes across a landscape. In this sense, the importance of ecological sciences in sand-control programs is not that they introduce the livelinesses of biology as a corrective to the inorganicity of geophysics but, rather, that they can elicit modes of futurity from sand that allow state sand engineers to expand the chronopolitical field in which they operate. Ecology and other sciences in the interdisciplinary apparatus of desert engineering might in this way be approached as attempts at investing and unearthing what Aihwa Ong, borrowing from her bioscience interlocutors, calls pluripotency. They aim to use sand to shape futures against the singular time of encroachment, effecting a "movement from the actual to the virtual, from singularity to multiplicity" (Ong 2016, 12). As a competitive mode of sand-time, this emphasis on succession allows dunes to exist perpetually at a crossing point between two sand time lines: the mobility of encroaching dunes contends with ecological succession, wherein sand becomes self-stabilizing botanical substrate.

In ecological succession, a bare dune is a starting point for a new process whose futures are already contained in the first weeds on a dune's slope. Teacher Xia, an ecologist a few years Li Ming's senior, insists that dunes are sites of ecological potential. She sinks a shovel and digs a hole less than a meter into the side of a dune to expose damp sand. "Look—sand is excellent for holding water and preventing desertification. These dunes can support robust plant life," she insists, pointing to the dunescape as a massive reservoir for potential botanical stabilization projects. Xia's desert, is, following Elizabeth Povinelli, a "space where life was, is not now, or could provide the conditions for life" (2016, 170). It is the habitat for a forest to come, which would realize the potential ecology of sand as a weapon against sand's volatile geophysics. A water-sealing dune, she insists, is a potential habitat for sand-engineering plants and for a scientific forestry that adheres to and mimics the principles of ecological succession.

PIONEERS AND KICK-STARTERS

A restless dune is also a point of eco-geophysical intervention. It shifts from the result of desertification to a cusp between two modes of environmental time. At this cusp, the future is a site of technical intervention where "the sense of the present as ruined time" (Scott 2014, 12) is also the precondition for sand ecology to begin to root stability back into the land. Dune sands appear as both an aftermath of degradation and a growth medium for tenacious infrastructuralizing species. Ecological sciences assemble into a political toolkit that collects engineering techniques as tactical means of emplotting and realizing other futures. They aim to forcefully derail sand-time from one track to another, one where the desert is a future and not a future burial. It is a means of effecting a kind of temporal capture, whereby the beginning of the succession process already gestures at its completion.

It should be no surprise then, that although the stability of an ecological community is a goal, it is the pioneer species and not the climax community, the beginnings and ends of the successional regime, that student scientists are trained to notice. On an early morning dune surveying trip, recent PhD and specialist in remote sensing Liu Jun leads a van full of students several dunes deep into Naiman's Ke'erqin Sandy Lands. We follow him in a loose cluster in the depressions between the high slopes of dunes until he signals us to stop. He reaches down to pull up a tiny green sprig of a leaf, about the size of a clover. It pulls out easily of the loose sand, and several grains of sand hang delicately on its tender root.

"This is *shami* [*Agriophyllum squarrosum*]," he says, a tenacious grass adapted to dry and unstructured sand. Looking around now, I notice that there is a sparse carpet of *shami* on many parts of the dune. I am not the only one who hasn't noticed until now. "*Shami* grows quickly, and it is one of a few plants that thrives on mobile dunes," he says. "It may look like nothing, but *shami* is centrally important in allowing other plants to establish." Pioneer species like *shami* are crucial not only because they are proof of ecological process but precisely because they also change the soil conditions in ways that allow other plants to establish. In the sparse leaves and blades, it is possible to envision, with some imagination, shrubs and trees growing on the surface of a stabilized dune.

Pioneer species, then, kick-start ecological time, promoting subsequent waves of colonization. They matter especially because they indicate the renewal of an ecological time already occupied by its projected realization. Many ecologists at the station study pioneer species. Li Ming and a team of ecologists are especially interested in studying a shrub, *yanhao* (*Artemisia halodendron*), to understand how it prepares sand as a habitat for larger flora. *Yanhao* can continue to grow on fixed dunes and therefore limns stages of succession. A study published in 2014, coming out of research at Naiman, argues that *yanhao* "under nutrient limitation [in shifting dunes] is more likely to manage with a low level of nutrients in senescing leaves, giving this species an advantage in infertile soil" (Li et al. 2014, 182). In particular, they suggest that in nutrient-poor environments, the shrub incompletely reabsorbs foliar nutrients, which means that the plant and its falling leaves contribute "to the return of high-quality litter to the soil."

This starts a multiplier effect by which decomposing plant matter "accelerates leaf litter decomposition and nutrient mineralization" (183). In their conclusions, they suggest that *A. halodendron* accelerates the process of dune restoration. The plant itself here appears as both an indicator and an accelerator of ecological time on the dune, as it speeds the process of succession by stimulating the improvement of the dune's nutrient environment for other plants. The manipulation and acceleration of successive communities of plants appear as part of a potential armory of sand-binding interventions, an engineering technique alongside and in tandem with infrastructures like grids. Indeed, ecologists discuss the straw and nylon grids as ways of engineering habitats for sand-pioneering species by creating tiny wind shelters where shrubs and grasses can more likely root.

As we sit on the institute's watchtower, we face the sunset, looking where the open dunescape breaks into a patchwork of grasses and pioneer shrubs. Little Tu leans back, as if to linger for a moment in the quiet calm of a future always starting anew. For the ecologists at Naiman, if dunes could be left alone and pioneer species could be allowed to establish, then ecology would take its course. Where much of contemporary ecology outside of China has shifted from the orderliness of succession to chaos ecologies (Worster 1990), these ecologists see the dunes as the site of a latent ecology just about to burst back into life. For them, ecological time propagates into a politics, wherein succession is the template of an ongoing reactivation of time; its stagelike progressivism echoes, formally, that of Chinese socialism's theory of itself.[17]

As those most invested in sand control seek to contain it, temporalities of sand figure in diverse temporal formations. In this way, sand shifts subtly from an environmental problem open to pacification by technopolitical intervention, to become a key actant in shaping the governmental interventions meant to finally control it. In many modes of encounter, more-than-human landscapes scaffold emplotments of political and environmental futurity that relay through more-than-human becomings. Sand becomes the material condition of a metahistorical form that emerges in complex physical, political, and ecological intra-action, even as it shapes new forms of political time.

In the worlds of sand control, the horizons of politics take their shape through the horizons of the expanding deserts, beyond the timelessness of Nature and the imminent endings of disaster environmentalisms. Vibrant, potent, and irreducible to human design, environmental materialities and processes displace and force open an anthropocentric temporal imagination. At the shores of the desert, we may yet discern in the sand ways of living at the end of the world, enduring its erosions or opening the future anew.

PART II   Fine Particulate Matter

How I wanted to be that sky—to hold every flying and falling at once.

Ocean Vuong, "On Earth We're Briefly Gorgeous"

4 Particulate Exposures

BECOMING-ATMOSPHERIC IN PM2.5

The U.S. AQI does not include recommendations for
PM2.5 levels above 500, but levels are sometimes worse
("beyond index"). What should I do?
  • Pollution is hazardous at these levels. Everyone
    should take steps to reduce their exposure when
    particle pollution levels are in this range.
  • Staying indoors—in a room or building with filtered
    air—and reducing your activity levels are the best ways
    to reduce the amount of particle pollution you breathe
    into your lungs. Read on for more information on
    steps to help reduce your exposure to short episodes
    of high levels of PM2.5.
  • Links to recommendations for reducing exposure to
    smoke from fires are available below. These recom-
    mendations may help reduce exposure during short-
    term pollution episodes in which PM2.5 levels are
    above 500, since fine particles (PM2.5) are the
    primary pollutant in wildfire smoke.

US Embassy, Beijing

On a brisk morning in the dead of Beijing's long winter, Xi Jinping
went outside. It was the fourth of a dayslong bout of impenetrable haze
(*wumai*), just upgraded by state meteorology agencies from a yellow air
quality alert to a hazardous orange.[1] The February 2014 upgrade corre-
sponded to off-scale PM2.5 readings, as reported by the US Embassy. In
carefully planned spontaneity, the highest official in China allowed himself

Figure 13. Xi Jinping breathes without a mask.
Photo from *China Daily.*

to be photographed throwing in his lot with the common Beijinger. With
hefty entourage and press in tow, he ate a steamer of meat buns in a local
shop and strolled through the Nanluoguxiang neighborhood. Smiling, he
played Uncle Xi, *Xi Dada*, the avuncular leader. He asked after residents
and shook hands as he wandered through Beijing's iconic *hutong* alleyways, gentrified into a tourist pedestrian zone. The entourage allowed his
picture to be taken by passersby, who quickly uploaded images to Chinese
social media (fig. 13).

As shown in the epigraph above, the US Embassy in Beijing, in mis-
sives available to American nationals in China, offered safety recommendations that posed the threat of the city's atmosphere to human bodies
through analogy to wildfire smoke.[2] But state media outlets covered Xi's
brief visit breathlessly. They waxed poetic on the "residual warmth" of his
presence in the bun shop. In the opinion columns of the official press,
self-proclaimed observers of the pulse of public opinion described every
detail of his bearing. From the plainness of his speech to his insistence
on "wearing the clothes of the common people" (*"chuan baixing zhi yi"*)
(Wang 2014), their panegyrics fawn over the president's *qinmin*, his tender intimacy with the People. Haze offered an opportunity for shoring up
the president's populist credentials. Fine particulate matter refracts light
into a softly glowing white. Four days of state silence into the haze event,
a PR blitz in bad air. Carefully selected "found" cell phone photography
drummed up a viral event: the smiling president wading through a pool
of admirers and handlers, all shrouded in the soft halo of radiant haze.

In covering the visit, then covering it again as an apparent viral sensa-
tion, every state media account fixates on Xi's face. It is conspicuously
bare. He wears no face mask to separate himself from public air. The
official press interprets this muscular act of exposure as evidence for
the ordinariness of Xi's political style. Official commentators offer the

absence of a filter mask as a mode of sartorial solidarity with the country's breathing masses.[3] The country's highest official was throwing in his lot with common folk by exposing himself to the ambient damages of unprotected breath. "Xi Jinping did not turn away, nor did he hide from it. Indeed, he was with the masses (*minzhong*), together breathing unclean air" (Wang 2014). Newspapers captioned the photographs with a ready-made slogan in the rarefied style of a six-character couplet in officialese: "Breathing together, sharing a common destiny" (*tong huxi, gong mingyun*).

The scene foregrounds the president's breath as an act of empathetic exposure.[4] Its quick congealing into the couplet-slogan is a lightly threatening primer on new rhetorical conventions for the description of urban air.[5] Their syntactic paralleling, with its overcompensating optimism, is a set of guardrails demarcating the rhetorical limits of how bad air can be discussed. The slogan's demand for togetherness, for sharing, for that which is common, is a warning against other atmospheric collectives that might yet be breathing themselves into shape.

This acknowledgment of shared respiratory vulnerability attempts to redirect concern over the catastrophic air pollution into a focus on the president's body, as well as to offer a fixed language for containing the potential explosiveness of atmosphere as a political substance. The spectacle of unmasked breathing, as the slogan and its repetition emphasize, draws the potentate and the plebe forcibly together in that destiny as sharers of suffering. The air has shifted into another, the medium through which the Party-State's body politic confirms itself again: the Party and the People are one. This "state romance," wherein breathers are forced by the political symbology of Xi's exposed face demands "living happily ever after" (Shange 2019, 4) with the asthmatic state.

The subtle coercion of this romance is quite literally a toxic relationship. The possibility of grievance is undercut through a subtle threat that togetherness and sharing are demands to identify with the state apparatus as a suffering Chinese body. Following Dai Jinhua, breathers "are forced to share in the government's so-called hardship" (2018, 7). Xi has reconstituted the particulate fellowship of embattled urban breathing into a demand for public sympathy for bemasked officials.

In a moment defined through yawning inequalities and social fractures, "Breathing Together, Sharing a Common Destiny," sublimates public outrage over poor air quality into a cheerful threat that the air will not be tolerated as an occasion of political debate. But by the 2010s, it had become one of the most explosive political questions in China, aired out in multiple domains of public life and culture. That breathing could become a state-fomented viral media event emphasizes the willful exposure of Xi's presidential lungs to dense particulate matter as political performance. This performance seeks to neutralize exposure and the capacity of microscopic aerosol particles to challenge the political status quo into an ambient non-event. Bad air was to be a new generalized condition of Chinese state life: the administration recognizes it and is working on it.

The ordinary act of breathing here appears as a visceral erosion, and so Xi's masklessness is a spectacle of vulnerability that aims to cement his identification with people, as much as his ordinary language and clothing, as much as his gusto in eating meat buns. Onlookers inducted into *breathing together* are enjoined to understand their own erosion by breathing as a mode of empathy for the president, whose defiant act of breath is a respiratory sacrifice and act of solidarity. In Xi's naked face, splashing across China's newsfeeds, national unity appears as a respiratory principle. In the conspiracy of breath and fine particulate matter, of survival and injury, bad air is an architecture of *sharing a common destiny*. The *together* in *breathing together* takes on the quality of an implicit threat: submit to *this* formatting of shared damages. Speak and breathe in *this* genre of forced solidarity.

Nonetheless, Xi's decision to appear without a mask draws symbolic weight from the implicit understanding that it is indeed a decision. For those in power, whose cyborg respiration is achieved with an arsenal of masks, air purifiers, and engineered airspaces, sharing in common air is a certain kind of choice. Here, the language of the masses and the stirring format of the slogan invokes a kind of throwback socialism. This idealized socialism, rewrought as state spin on bad air that it is widely held responsible for failing to resolve, demands that the Party and the People are one. This mobilization of a socialist People seeks to reanimate a teleology of shared struggle and eventual triumph against the air: it seeks to domesticate the wildness of Beijing's meteorology into the futures of revived Revolution.

But after decades of Reform and Opening and the proliferation of new subjectivities (Rofel 2007), the People and its revolutionary sociology of masses also return as a historically charged container for capturing and constraining any emergent breathing public that might take shape with the dynamics of modern weather. The slogan and its cheerful-threatening media blitz are salvos against the formation, in the play of particulates and respiratory systems, of a mass breathing subject. Breathing as a meteorological basis of political collectivity is a site of potential friction with the Party apparatus, highlighting its slipping monopoly on the ability to define The People. The recirculation of socialist tropes around Xi's face, especially the notion of the masses (*qunzhong*), in this regard, must be understood as an attempt to contain the explosive political potentials of bad air in the preemptive revival of a charismatic People.

The president's face, presented with nervous gravitas in state press, is an attempt to contain and channel how air pollution can matter. In it, the contest plays out over the embodiment of modern weather in the Chinese city—a tension between "new forms-of-life and forms-of-death—both existing and imagined" (Bernstein 2019, 10). Unmasked, it forces breathing publics into conspiracy with the most powerful as sharers in respiratory suffering. *Breathing together*, breathing-with, *con-spirare*. It acknowledges necrotic conspiracy—a condition of life in the air in which breathing together also means decaying together—while carefully shaping how such decay can matter. This political capture of breathing is a stymying demand for an appreciation of the difficult predicament of the most powerful, rather than an occasion for demanding the state's protection of breathers. There is a subtle chide in this victimology of the powerful, as if ordinary people did not have enough empathy for the plight of their leaders, sharing, as they are, in the debilitations of breath.[6]

But aerosols do not blow through halls of power or settle in the lungs of powerful men in the same ways as they do in the multitude of compromised breathing spaces that make up the city and its tableau of compromised life-in-air. "Breathing together rarely means breathing the same" (Choy 2016). Uncontained by the political designs that pass through the president's bared and bated breath, other conspiracies may take shape, and other ways of breathing and dying together may be possible.

———

This chapter explores the proliferation of experimental figurations of breathing bodies and modes of political collectivity in Beijing. It attends to shifting figurations of body, collective, atmosphere, and state through breathing, in its reconfiguration in particulate matter politics, as the condition of a necrotic relation to China's meteorological contemporary. I trace these reconfigurations through considering public culture and explosive debates over the technics and standards of air quality measurement in Beijing. By centering breathing (see Škof and Berndtson 2018) as an interface with meteorological change, it explores reconfigurations of bodies, atmospheres, technosciences, and collectives. What could sociality be if it began in the thick of a meteorological condition, if it were a practice of weather-worlding? If it took shape like clouds as well as with slogans, if it rose with dust as well as with revolution? What is at stake in such questions is locating the becoming-meteorological of political collectivity in the Chinese capital through a series of provisional respiratory formations that disorient *both* the liberal expectation of a much-anticipated civil society in China *and* the revival of an earlier figuration of The People.

The chapter takes its cues from speculative fiction, social media campaigns, and the political theatrics of haze, each of which offer resources for attending to particulate exposure as a laboratory of modes of political embodiment and disembodiment that stand in a series of oblique relations to the putative stability of the demand of Xi's administration for shared breath and destiny. As the meteorological dynamics of dusty air have unfolded unexpected geometries of the body and body politic, so, too, has urban air become multiply enacted as a terrain of political maneuver. We orient through a succession of scenes of debilitated breath to attend to how the separated of any enclosed body is unfolded through atmospheric relation. Life and breath orient us toward the mass breathing publics that the circulating image of Xi's unmasked face seeks to exorcise.

"Where is the human body if it is viewed from the lung" (Povinelli 2016, 42)? In 2013, Beijing municipal health bureaus reported a near 50 percent increase in incidence of lung cancers in the span of a decade. A study in the same year reported that since 1981, at the early outset of the massive experiments of Reform and Opening, across northern China, inhalation of suspended particulates had lowered average life expectancy in the region by more than five years. It quantified this damage as a loss of an aggregate 2.5 billion human life-years (Chen et al. 2013). This macabre calculus renders the relationship between life and breath as a series of double binds.

This kind of breathing, measured in lost life-years, makes *life* and *death* loose synonyms; the vital functions that sustain life appear as inseparable from their own undoing.

Such studies recall an everyday complaint of life in the Chinese capital, where Beijingers narrate a necro-respiratory economy in which the conditions of living are inseparable from the atmospheric etiologies of premature death. In this light, Xi's walk must be understood not as a tightening authoritarian line on the politics of air pollution, but as a provisional attempt by state media to conjure a late socialist social that might supplant and defuse the destabilizing potentials of the endangered breath that has become synonymous with city life.[7] Four days into the current haze and several years in the making after public outcry over air quality, Xi's walk has tacitly acknowledged that the air is not only a political substance, but also the medium through which a specifically meteorological kind of sociality of breath is to be traced and contested.

The media spectacle of Xi's ordinary breath thus must be understood in relation to the latent breathing collectivities that it seeks to preempt. It proceeds by shoehorning particulate exposures and the profoundly complex geochemistry and physics of the atmosphere into socialist frames of shared struggle. But as a phantasmatic legitimating figure of struggle shifts phases into matters of the distribution of aerosol damages through urban breathers, late socialism reappears as a general condition of respiratory distress.[8]

Each of these versions of collectivity, up to and including the *common destiny* presaged in Xi's uncovered face, is triangulated in a necrotic knot. We attend, then, not to the cultural politics of air pollution, but to the experimental meteorology of massively distributed decay as a medium of tentative political emergence. This experimental meteorology of politics can be traced ethnographically through the multiple permutations of breathing flesh, body-penetrating particulate matter, and modes of political relation and alienation conditioned by China's meteorological contemporary. Crucially, each scene in this chapter poses civic life as a kind of collective but unevenly distributed necrosis. I like this word, *necrosis*, because it captures something of the sense, sometimes latent and sometimes spectacular, that "something in Beijing's air is slowly killing people" (Kay 2020, 25). It centers the erosion of vital processes that is forefronted when we consider lung cancer statistics and the diminishment of aggregate life-years as part of a Chinese weather report, offering a way of staying close to bodies, gathered and dispersed by air, without forcing a hard distinction between life and death. Necrosis allows for living to also be dying, ensnaring us into the fraught calculus of life in a city where the continuation of the body is also its gradual undoing. Necrosis is in the decay, the erosion, and deathliness that is concurrent with living. In necrosis, life and death are a question of gradients and simultaneities, of living as a mode of dying.

This condition is one of particulate exposure—a "contact between flesh and misplaced matter" (Mitman, Murphy, and Sellers 2004, 13) oriented toward inhaled aerosols as a medium of shared and divergent atmospheric vulnerabilities. Exposure, writes Stacy Alaimo, offers a site for formulating "modes of environmental ethics and politics that may be penetrated with feminist histories" (2016, 94), especially those in which the modernist and masculinist injunction to bodily impermeability is faced with "an openness to the material world in which we are immersed" (2016, 91). This condition of exposure is non-innocent, political, and also existential. In substantiating exposure as a particulate phenomenon that destabilizes the coherence of the boundary between living in dying into permutations of necrosis, here exposure becomes a site of political experiment at the interfaces of bodies, states, and weather systems.

Exposure, so wrought, evinces oblique possibilities of collectivity. *Breathing together* is an open question about the forms that conspiracy can take in haze. This is clearest in the revelation of breath in stunning particulate density as not only a site of biometeorological derangement, but also as a material site through which to consider atmopolitical formations in the making. The conspiracies that concern us rock both incantatory revivals of the socialist mass and the white-knuckled expectation of liberal democratic uprising off their already uncertain footing.

In what follows, I offer a recent history of living and dying in the air, beginning several years before Xi's maskless walk in 2014, when public debate in Beijing and elsewhere condensed around arcane details of air quality measurement and reporting. In 2011, dramas over PM2.5, a measure of fine particulate matter, played out unexpectedly in a diplomatic fracas between the Chinese Ministry of Foreign Affairs and the US

Embassy in Beijing, and then, improbably, on the massive social media platforms of a firewall-jumping Chinese tycoon. I show that what is at stake in this specific measure was not only data transparency, which foreign commentators roundly assessed as a long-awaited proof of the resurgence of a proto-democratic movement. Rather, at the center of this was the attempt to compose new images of meteorological corporeality and exposure through lung-penetrating PM2.5 as precisely the intersection of medical and environmental domains. PM2.5, that is, was not simply a demand for more and better data, but an attempt at reformatting body and atmosphere to concretize a public sense of dying-in-air.

In the next section, I explore how breathing emerges as the framework for an emergent form of political embodiment, one that by 2014, the state attempted to domesticate into a respiratory socialist solidarity that would amplify the powers of the state through the air rather than erode them. In the public outcry demanding PM2.5 measurement we witness the explicit theorization of the body in the relentless incapacity to not breathe, the autonomic functions of bodily maintenance assemble the body dangerously in self-negating relation with airs and others. I suggest that such embodiment focuses us not on the agency of political collectives, but rather to an attention of double negative conditions, where breathing is not a doing or an action as much as it is a not-being-able-to-not-do. This is an incitement to a less triumphal form of politics, a becoming-atmospheric of the body politic into a necrotic public. This public appears by disappearing in images of choking and compromised life.

But let us begin with a story. Chen Qiufan's 2015 short story "The Smog Society" traces the formation and dissolution of a citizen science organization in an unnamed Chinese city.[9] The endless haze of a sky "pouring down dust endlessly" has stranded the city's remaining populace, while the wealthiest have evacuated to clearer skies or stay, barricading themselves behind expensive filter masks. Members of the group, popularly styled The Smog Society, quietly collect air quality data, mostly without shared or stated aim besides breaking up the enervating monotony of life in the dust-cocooned city. Lao Sun, the story's smog-depressed protagonist, illustrates the political ambivalence of the group. "The Smog Society wasn't as radical as some green groups, but it wasn't the government's cheering squad either." Lao Sun continues to collect data, unable to clearly formulate a motivation or theorize the Society's work. In the climactic scene, the group's data reveals a causal relation between civic mood and civic weather: depression and smog cause each other. With this revelation the loosely organized Society is forced to quietly disperse under tacit pressure from government officials. "Smog buddies," the organization's members, learn to unsee each other, passing each other in the streets as strangers. The Smog Society dissipates like the haze that draws its members together.

In the story, smog is a society. Chen's account of the Society is not about the principled self-organization of informed citizens into a public against state and smog. Rather, Chen poses smog and society as conjoined modes of relation. The coalescing and then dispersal of the Society mirrors that of its Smog. Chen traces their mirrored modes of relation in the feedback of social and particulate compositions. Smog and Society take the shape of one another in a choreography across phases and domains. At the same time, the corporeality of individual bodies is *part* of the smog's expansive and penetrating particulate reach. "Even with the filter mask," Chen writes, "you felt as if the smog could worm its way through everything . . . stuff your chest full until you couldn't breathe; and turn your brain into a drum of concrete too thick to stir." The ability to adjust the body's exposure is classed—the rich have moved away, the well-heeled remainers disappear behind ever more elaborate membranes and masks. This collective embodiment through particulates climaxes in the literal incarnation of smog as an animal enfleshment, a body studded with human bodies. "People were like parasites burrowed into the smog" (2015). In the story, smog itself—and not shared political conviction or the human bonds of family or friendship—is a shifty physical and affective scaffold that holds suffering bodies together in temporary formation.

"The Smog Society" and its eponymous organization is not a story of heroic atmospheric resistance nor a lament over the failures of that resistance. In the haze, the lines that structure a state-society relation diffract into a scatter of vague and intermediate disorientations. Social and meteorological feedbacks and changes, in the story, make smog and society part of the unfurling of a broader weather pattern. The Society has an episodic temporality; it gains coherence as particulate densities and exposures condense. Like smog flushed out of a city basin by a rush of wind, it simply falls apart as meteorological and political dynamics shift. Collective formations and haze, in Chen's story do not *cause* each other in a unidirectional way. Rather, they mirror each other, a point made literal in the story's climactic revelation. Collective human psychological states, the Society establishes, generate "large-scale bioelectric fields" that "affec[t] the distribution of aerosolized particles," making the smog an affective atmosphere in a literal way as depression and aerosols affect each other. The Smog in Smog Society multiplies: it describes the boundary object that binds members, but also describes the continuity of social and meteorological formations.

The Smog Society and Xi's breathing masses fold disparate socialities out of the air, reformatting the geometry of the social with aerosol morphologies. Both demonstrate that "unclean air" is more than an occasion for political attention, more than a passive object on which power acts. Chen teaches me to attend to bodies, affects, and Societies, as nodes in complex meteorological systems. Building on these insights—that sociality might be tracked as a process of expansive meteorological emergence and dissolution, rather than through the tracing of ontologically prior body-subjects—our attention condenses with the political worlds that are taking shape and vying with one another through Beijing's air conditions.

Following Chen, we attend to collectivities that take shape and break in haze events not as matters of self-organizing human political forms, but as more-than-human formations in a poetics of particulate matter. Vulnerability and ailments, like breathing, rather than appearing as damages contained in individual bodies, become entailments of relation in and through atmospheric medium. If "bodies are stew pots cooking up a new form of posthuman politics with new forms of posthuman corporealities" (Povinelli 2017, 509), breathing situates these forms of corporeality in an unending environmental process. It sets for us the task of tracking forms of sociality, human and more, that eschew lists of stock political characters and stances inherited from either liberal or socialist orthodoxy. Moving across particulate gradients, airways, and the membrane of the lung allows us to point our attention instead toward the autonomic functions of vital persistence and decay. We wander in the diffuse gunks of atmospheric interrelation.

2.5 MICRONS

How will I know when conditions are better?

- Air quality conditions in Beijing can change
  rapidly. Check the U.S. Embassy or Chinese government
  air quality monitoring web pages for the most recent
  hourly PM2.5 readings. These readings can help you
  determine when to take steps to reduce your exposure.
- Also pay attention to weather forecasts; these can help
  you plan your activities for times when air quality
  improves, such as when winds are forecast that clear the
  air. When the air clears, and AQI readings are low, take
  advantage of these times to get outdoors.

Beijing, December 2011: coming out of the winter gray into a restaurant, my friend Li Qiang apologizes through a cough for his lateness. Air purifiers hum ineffectually. We talk, like everyone else, of the gray pall that hangs heavily on the city, and, in these early days, over our conversation. Days of horrendous air quality had contracted daytime visibility to dozens of meters and grounded upwards of seven hundred planes at Beijing Capital Airport, purpose-built just three years earlier for the Beijing Olympics. In days, this bout of pollution, diffuse and ubiquitous, would be christened in the foreign press as the first of many Airpocalypses, linking darkened sky to invocations of the end of times.

Li Qiang has been in Beijing for five years since leaving his home in the frigid northeast for university. He has stayed on to work as an environmental journalist, but announces each time I see him that he is going to leave soon, abroad maybe, where the political and physical atmospheres may allow for more breathing space. He senses in the air a diffuse menace, his own prospects for upward mobility in the city contracting like the stunted visibility of a storm. Each year he lives and breathes in the city, he jokes, he trades for a year at the end of his life. He hopes that after I leave China, we will meet in America, or Germany, if he can find a fellowship.

It is not the first haze in the city, and not even the first this year. Like other friends in the city, he sometimes greets me when we meet with numbers reposted onto Chinese Weibo micro-blogging accounts from the US

Embassy's Twitter feed, which releases hourly PM2.5 readings. On his phone, he swipes through columns of numbers, each a measure of the same air, including both Chinese and US air quality indexes, but also different measures of particulate matter concentration. I have thus far refused to get a smartphone and don't quite understand what apps are. Li Qiang, playfully exasperated that he needs to educate a visiting foreigner on technology, teaches me that what stands out across columns are the stark discrepancies in measures of the *same* air. As Tim Choy shows, air quality indices and reporting schemes vary across jurisdictions. "The standards for danger are different in different places" (2011, 163). There are stark variations across numbers reported by the Beijing meteorological bureau and by the US Embassy, as well as their appraisal of the situation: a Chinese "moderate" seems to correspond to an American "hazardous."

On days like this, it is customary to complain about the air and the government. Where a few years earlier, locals adopted a certain cavalier attitude toward the air, boasting iron lungs among the other assets of Beijingers, this bravado has evaporated into the haze. In 2008, in the prelude to the Olympics, to criticize the air was to criticize the entire program of national re-arrival of which the Olympics was a final proof. I laugh as Li Qiang puzzles through the numbers. With his mask on the restaurant table, he smokes cigarette after cigarette as we sip on weak lukewarm Chinese beer, shrugging. "It's nothing worse than what you breathe on your walk home or even in your crappy apartment," he chides. What could the difference between conscious and inadvertent inhalation of particles be, besides a small moment of pleasure? It is a complaint constantly voiced in the city, an echo of a joke circulating Chinese social media in these haziest of days: "Do you know what welfare benefits you get by living in the capital? All you need to do is open your windows for a smoke!" The joke frames consumer choice in Reform as a supermarket of debilitations. With a shrug, he says he may quit when he moves.

Different measures of particulate density also appear to suggest radically different atmospheric conditions. Official meteorology registers the particulate concentration in PM10, particles measured at 10 microns, as merely "moderate," and in the days of haze, the year-end meteorological recaps on state-run television chirpily report an annual increase in "blue-sky days," a nebulous measure of official air quality that municipalities use to affirm attainment of a central air-quality standard. Li Qiang and other friends in Beijing are in the habit of assuming that state weather reports, with their rosy announcements of year-on-year improvement, are flat lies. This is a suspicion that US Embassy numbers seem to confirm. "They can, without a change in facial expression, tell us on a day like this that the sky is blue!"

@BEIJINGAIR

Six months before the 2008 Beijing Olympic Games, the US Embassy in Beijing installed a MetOne BAM 1020 continuous particulate monitor on the roof of the diplomatic compound. The official rationale for the installation of the monitoring device was to provide adequate air pollution data for embassy staff and the influx of American visitors during the "Green" Summer Olympics and then for the benefit of Americans in Beijing after the Games. The Embassy also made available lists of recommendations and frequently asked questions and answers about limiting exposure to fine particulate matter, in English, for US citizens. Beginning in July of 2008 readings from the monitor were, and continue to be, automatically released on an hourly basis through a Twitter account, @BeijingAir, operated by the Embassy. Each automatically generated tweet reports PM2.5—a much finer measure of particulate matter than the PM10 that was the standard in Chinese reports of local air—as well as ozone, and an air quality rating, calculated to the standard of the US EPA. Since then, an Ecotech EC9810 monitor has also been installed.

In principle, because Twitter is officially blocked in China, this data feed operated in a digital gray zone. Chinese internet censorship offered a protracted space of plausible deniability for Embassy staff because the tweets were not publicly available to Chinese nationals without specialized, but relatively easily obtained, software for breaching state firewalls. These numbers were thus officially invisible in China and yet easily accessible, which had the effect of making air quality measurement and reporting a potential diplomatic snafu, one that would touch on other verboten topics such as state internet regulations and censorship, as well as the more arcane technical questions of how air quality was measured or calculated: where were the monitoring devices? At what altitude? Is it true that they would be sprayed down before the numbers were reported? In June of 2009, *Time magazine* published a story about the monitors and @BeijingAir (Ramzy 2009), which drove a spike in followers, which, according to unclassified diplomatic cables released by WikiLeaks, were mostly firewall-jumping Chinese users.

According to the same cable, the Chinese Ministry of Foreign Affairs requested a meeting at the Embassy on July 7, 2009. The representative of the Ministry's Office of US Affairs complained about the data being released, "which in their view 'conflict[ed]' with 'official' data posted by the Beijing [Environment Protection Bureau]" (Embassy Beijing 2009). In addition to the indignity of a foreign government measuring domestic air—an apparent infringement of meteorological sovereignty—the Chinese official argued that the very existence of the Twitter feed sowed confusion and caused undesirable "'social consequences' among the Chinese public." It urged a compromise that might limit the Twitter accounts followers to US nationals traveling and residing in Beijing—a redundant request, given that Twitter was already officially inaccessible to Chinese nationals. The US government does not appear to have complied, and soon accounts popped up on Weibo, a megapopular microblogging platform in China, that reposted @BeijingAir tweets and juxtaposed them with Chinese appraisals of the same air.

Besides the attempt to maintain a monopoly on meteorological monitoring and reporting, much of the conflict between the Ministry of Foreign Affairs and the US Embassy appears to be over technical discrepancies in how air quality and especially particulate matter would be characterized, and the format in which that information would be disseminated. The US Embassy, for instance, reported in real-time while the Beijing Environmental Protection Bureau offered after-the-fact recaps for the previous twenty-four-hour period. That its tweets released PM2.5 readings, which consistently characterized air as more hazardous than the Environmental Protection Bureau's PM10, was identified in the cable as an important difference: "Beijing EPB currently has the capacity to collect PM 2.5 data, but the agency only chooses to make available to the public data on PM 10 (10 micrometers in diameter), and links the PM 10 data to Beijing EPB's own 'Air Pollution Index (API)' definitions" (Embassy Beijing 2009).

The impropriety that the Ministry of Foreign Affairs accused the US Embassy of in measuring and reporting Chinese air in conflict with Chinese meteorological reports became an issue of contention between the governments. The US Embassy and the Chinese Ministry of Foreign Affairs debated over the appropriateness of American reports of Chinese air. This was largely interpreted by representatives of the Chinese government as an insulting infringement of the sovereignty of the Chinese government, who argued that "Beijing [Environmental Protection Bureau] should be the sole authoritative voice for making pronouncements on Beijing's air quality" (Cable 09BEIJING 1945_a). Nonetheless American reporting of PM2.5 through @BeijingAir continued and continues, substantiating Beijing's atmosphere as a collision between datasets and technical standards.

But technically, the reporting and measurement conventions differed, and this was the crux of both the debate over transparency in environmental reporting and the *kind* of atmosphere and the kind of body-atmospheric relation that was deemed important to be known. The numbers differed wildly also because of the scale at which the governments measured particulates and calculated their Air Quality Indexes (AQI), as well as significant disjuncture over the points at which grades of air quality were marked. While Chinese meteorological agencies reported concentration of PM10 particulates, at 10 microns, US embassy reports publicize concentration levels of the much finer lung-penetrating PM2.5, 2.5 micrometers, a measure of the tiniest ambient particles that thwart the body's filtering mechanisms and penetrate deep into vulnerable urban lungs and bloodstreams. This difference in PM measurements, indeed, was consistently expressed in public culture and social media, as well as on state news, in terms of the penetrability of lungs, and in the difference between particulate matter as a broadly environmental question and as a medical question of environmental exposure.

In the winter of 2011, as the stark difference between official and US measures, PM10 and PM2.5, met with a catastrophic haze event, these sets of scattered numbers, colors, and indexes, of "moderate" and "hazardous" ratings, began to find their way into more and more high-profile social media accounts, and then into public discourse. The difference between the two data sets, at its core, was not focused on the status of the Chinese state as a liar or manipulator of data—although certain claims have been made about the Chinese state moving its measuring stations farther and farther from the perpetual traffic jam of the central city (Andrews 2008). Li Qiang, for instance, already paid little attention to Chinese readings unless they were juxtaposed with other readings—he was attentive not to the data itself but to the gaps in comparison. He met the discovery that some measures of excellent air quality in China register as "unhealthy for sensitive groups" by US measures as a confirmation not of state lies, but as an almost parodic instance of the differentials in how the values of life and breath were to be calculated and evaluated.

Nor was it a matter of posing Chinese and American data sets against each other in terms of a battle between nations or political systems. What is at stake is not that US measurements were scientific and that Chinese ones were ideological. It was rather a technical matter that hinged on how air affects the body, hanging in the span of microns as a proxy for the penetrability of the respiratory system. The questions of diplomatic protocol and effacement of sovereignty that occupied the Chinese official at the US Embassy gave way to questions of how particulate matter data collected and reported at microscopically different scales offer different theories of the body as part of a continuum of meteorological exposures and infiltrations.

While the Chinese meteorological and foreign affairs authorities aimed to spin the matter of American reporting by mobilizing nationalist sentiment to dismiss American numbers, something else entirely was happening in social media and across the urban populace. The Embassy numbers drew the interest of Pan Shiyi, a billionaire real estate magnate who, like other mega-rich tycoons, had massive followings on social media. The numbers and standards took center stage, and while the Chinese government attempted to frame debate over the now widely known US Embassy particulate monitor as a problem of bilateral relations, for Pan, the conflict between the two sets of air quality data did not play out at the macro scale of international relations, but at the scale of the microscopic 7.5 microns between airborne particulates measured at 10 and 2.5.

In November of 2011, Pan Shiyi began reposting PM2.5 data released by the US Embassy, drawing attention to the difference between the two measurement standards to his more than seven million followers on

Weibo. The posts started a flame war on the platform. Du Shaozhong, head of the Beijing Environmental Protection Bureau, accused Pan of indulging in petty bickering (*koushui zhan*), to which Pan replied that he was not bickering with the state, but indeed expressing care and concern for "the elderly, the children, the families living in Beijing, in the living, vital environment (*shengcun huanjing*) of every Beijinger."[10] So provoked into a moral crusade for this living environment of airs and bodies, he began widely posting about the responsibility of the state. Rather than exhorting the government to resolve air pollution, he first demanded an immediate change to its air quality standard. The government, he posted, should begin posting PM2.5, which, he explained, "is more harmful to the human body. Beijing Environmental Protection already has PM2.5 data. . . . It should start making it public, shouldn't it?"[11]

Pan released on his Weibo feed a poll for his followers on whether or not they hoped the state would integrate PM2.5 into air quality standards. The poll ran:

> Experts say that PM2.5 air pollution is a great harm to the body. Only the national government can adopt mandatory standards which individual cities can then implement. Only knowing the seriousness of this problem can all people work with awareness to combat air pollution, changing their own unhealthy lifestyles and habits. Please invite your friends to vote and forward this poll. After a week, I will take the results and write a letter to the minister of the State Environmental Protection Department.

At the end of its week, the poll had racked up more than forty thousand votes. A full 98 percent of respondents supported the adjustment of air quality standards to include PM2.5 measurements, with more than 90 percent demanding it immediately. The poll and the campaign for PM2.5, which some commentators consider the most consequential social media campaign in Chinese history (*Economist* 2013), succeeded in making what just weeks before was an obscure minutia of air quality metrics into the protagonist in a sudden and ubiquitous pouring out of grievances over the air.

How are we to understand the charisma of PM2.5, which in the span of weeks had incited the air into discourse, and even direct grievances to the central government? The difference between PM10 and PM2.5 is not merely a question of the completeness of state measurement of the atmosphere or even a matter of an urban populace demanding transparent reporting. The 7.5 microns between them is a chasm between the two visions of the air as a political substance, hinging not on the atmosphere as an autonomous domain, but enacting it as a volatile mix of infiltrating particulate matters.

Pan glossed the difference between PM10 and PM2.5, importantly, on the different ways in which they imbricated bodies in the air, where the finer measurement allowed for the air to be discussed as a cause of bodily harm over prolonged exposure through breathing. In a retrospective on the campaign in 2014, Pan describes his own education on particulate measurement, spurred by a perplexity over the disparity in Chinese and US Embassy measurements. His moment of revelation occurred when he understood, in discussions with scientists and politicians, that "the smaller the particles in air pollution, the greater harm they are to human bodies. Larger particles [like PM10] enter the nose but are then filtered out and so cannot enter our bodies, but smaller particulate matter can enter the lungs, and even smaller particles can even move through the lungs into the bloodstream, causing very large harm to the body. This is the concept behind PM2.5."

It was the body's penetrability at stake, where breathing oscillated wildly between an innocuous vital process to a complicity in the relentless degradation of the body through its intake of particulates. If PM10 indicated a focus on what could generally be called the environment as an order of reality whose relation to human habitation is unclear, the feature of PM2.5 pollution that gained most traction with the Beijing populace was its capacity to afflict harm by entering the body through inhalation. PM2.5 gave a technical and political substantiation not only to the problem of the atmosphere, coming into view as a substance densely laden with suspended solids, but of the air in which one lived and had no choice but to breathe as the inescapable medium of a repetitive injury. Atmosphere and human bodies commingle in PM2.5 into a violent and shared continuum of "vital environment" (*shengcun huanjing*). And in this contact, limning dense atmosphere and the capillary surfaces of the body's deep interior, PM2.5 allowed for a transformation of air pollution as a question of detached "environmental" concern, measured dubiously by the blueness of the ashen sky, into the substance of a deeply embodied particulate exposure. It revealed the populace of Beijing not merely as bodies that could be injured, but as breathers.

Feeling the body, its life, and breath in PM2.5 is to witness its becoming-atmospheric in the technopolitics of discrepant datasets, and the sloshing of social media hype into public life. In conversations with friends during the dizzying unfurling of the atmospheric event and its smoggy socialities, the cascade of numbers had a pedagogical air to them, as if they might offer an atmo-etiological master key around which diffuse feelings might harden into suites of atmospheric symptoms. People I knew were training themselves to understand them, and thus to picture urban life as a breathing condition. This conspiracy played out as friends built a speculative symptomology of atmospheric states, learning to associate airport closures or a specific scratchiness in the throat to landmarks in the play of numbers, measures, and social media feeds. What do so many microns feel like slipping through the seams and rifts of a loosened mask? What does one make of that peculiar heaviness of becoming-atmospheric and of becoming an unwitting cement mixer in the same breath?

The demand for measurement of these particles precipitated a diffuse urban affect around the open secret of the city's environmental woes. PM2.5 provided a rhetorical anchor for the expression of grievances over air quality, which overwhelmingly turned on images of vulnerable breathing bodies. Prolix users, led by tycoons in the social media, were joined even by state-run media outlets that editorialized swift and satisfying government action lest popular discontent bubble over into more destabilizing effects. In mere weeks, over the 2012 Spring Festival holiday, the central government announced that it would begin collecting and disseminating PM2.5 data in major cities, a "New Year's Gift" to the people.

THE AUTONOMIC EVERY-BODY

> Who needs to take steps to reduce exposure when PM2.5 levels are "hazardous" or above on the AQI?
>   • Everyone needs to take steps to protect themselves when pollution levels are "hazardous" and above. Some

people are at higher risk from PM2.5 exposure. People
most at risk from particle pollution exposure include
those with heart or lung disease (including asthma and
chronic obstructive pulmonary disease-COPD), older
adults, and children. Research indicates that pregnant
women, newborns, and people with certain health
conditions, such as obesity or diabetes, also may be
more susceptible to PM-related effects.

Might exposure through the breath offer an anthropology of the body, and
a politics through environmental embodiment, an opening to the body
beyond its status as "cultural'"? Elias Canetti, writing of his friend, the
author Hermann Broch, asks, "Can we actually conceive of a literature that
stems from the experience of breathing?" (1979, 10) The very form of his
question hints at the uncanniness of an attention to breath, an experience
that is so automatic that naming it renders it strange to itself. Is breathing an act? What historical event sees breathing shift into the register of
"experience"? What has become of the breath, when "a sustained mind-
fulness of the air's breathability" (Sloterdijk 2009, 84) has supplanted its
autonomic repetition? Canetti's account vibrates with the aftermaths of
gas warfare in the Great War, where breathing makes bodies unwitting
accomplices in their own undoing. At its core, in these changed atmospheric times, breathing emerges as a "defenselessness," for "to nothing is
man more open than the air" (Canetti 1979).

   Urban coughs might be heard as the self-interpellation of a new site of
political-bodily investments. If Canetti considers a literature founded on
the breath, we might ask of an anthropology and a politics that likewise
"stems from the experience of breathing" in a particulate milieu where
breathing itself has become "an experience." But these possibilities must
be tracked through breathing itself as an autonomic process that runs
askance of a binary between agency and passivity. Breathing is neither
something that one does *per se*, which renders the "experience of breathing" a strangely discordant phrase, nor a passive act in the sense of a not-
doing. It is not synonymous with being acted upon. If breathing shapes
an emergent political actor, it does so through the various ways in which
breathing at once stages and then disrupts any binary of agency and its

negation, requiring a careful attention to breathing as it comes to matter, variously, in scenes of incapacitation and dispossession by exposure as a bodily and political condition that forces reflection in double negatives: breathing as what one cannot not do for long.

On days of hazardous air quality on the US scale, the US Embassy in Beijing advises "*everyone*" to take steps to protect themselves. Who, in bad air, does this "everyone" entail? In recent years, bouts of air pollution have been accompanied by flurries of air-related content in public culture. Thomas R. Johnson and Kathinka Fürst identify artistic responses to smog as a limited domain of activism in which visual representations of smog "can help generate public concern, which in turn can pressurize officials to take steps to address large-scale pollution" (2019). They argue that artistic takes on air pollution are a medium for disrupting official narratives concerning air pollution, which have attempted to naturalize it as a necessary effect of a fixed process of national economic development. In their analysis of a growing body of Chinese artists' responses to air pollution, they emphasize art as a mode of social engagement.

In doing so, they ask for an oblique approach to exploring air pollution as a political materiality, one that steers away from scenes of direct protest and dissent that align with agonistic liberal conceptions of the political. They focus on images of bodies, technologies, and the air itself, rendered shockingly visible as a palette of grays, off-whites, and ruddy sheens. In this section, I draw on this approach but shift my focus from auteurs to uncoordinated, public cultural responses to atmosphere as a problem for breathers. At stake in this is speculating over the possibilities of a body meteorologic, taking shape at the bodily atmospheric encounter of breath as an autonomic process. Collectivity, that is, appears not as an uprising against the state but rather the slow emergence of new modes of political relation through the political substance that air has become. This offers a chance to consider the cyborg images of particulate relation, a techno-bio-atmospheric poetics of particulate matter that propose modes of breathing together that shrug off state attempts to domesticate them.

"Chinese breathing" occupies a strange place in the anthropology of the body. In his seminal paper "Techniques of the Body," Marcel Mauss understood "breathing techniques in particular," central to the disciplinary and mystical practices of Daoism, to be "the basic aspect" in China (2007, 68).

In a reflection that founds the capture of "the body" as an anthropological object, attention to Chinese breathing bolsters Mauss's contention that "it is thanks to society that there is the certainty of pre-prepared movements, domination of the conscious over emotion and unconsciousness" (2007, 67). Chinese breathing for Mauss founds a more general subsumption of the body into the order of the social, the conversion of the most automatic of processes, through consciousness or culture, into anthropological objects. Margaret Lock and Judith Farquhar reflect on Chinese breathing disciplines to argue that "even the most foundational physiological processes can be subject to deeply formative training" (2007, 187). Chinese breath founds an anthropology of the body that transitions bare physiology into cultural fact, and into conscious manipulation. However breath, as an autonomic process, cannot be fully contained in Mauss's schema that transitions the body into culture through discipline and manipulation. The body's processes, that is, are not fully subsumed into the question of either their cultural representation or their domestication into the order of the social through "deeply formative training." What is at stake, then, is the possibility of picturing political futurities in which the collectives form and disperse like, and with, weather events, instead of rising into self-awareness and agency as the demands of History require.

Amid the increasing agitation for PM2.5 measurement, in haze events in the winter of 2011 and 2012, Chinese social media was flooded with images collected under the hashtag "#I don't want to be a human vacuum cleaner," (#我不要做人肉吸尘器). The quickly sedimented formal vernacular of these images was simple: they were facemask selfies, sometimes of individuals and sometimes of groups, with placards with the hashtag. They took place across Beijing, often in well-known city landmarks associated with state power like Tiananmen Square. The hashtag first appeared in 2011 and has periodically recurred with pollution events, creating a digital and visual archive of air pollution emergencies as an ever-growing mass of images of masked faces. The density of hashtagged images in social media thus mimics the density of particulate pollution, even as it links each of these events across years in the ever-increasing aggregation of timestamped images. The visual texture of particulate atmosphere persists while hairstyles, clothing, and indeed, mask styles change over time. As if linked by the same pollution event stretched over years, the atmosphere appears as a non-real-time assembly of bodies with faces obscured, increasing as the page scrolls.

The hashtag campaign aggregates images into "transient and critical gatherings" organized around the negative hope that one does not wish to become a vacuum cleaner. This mode of "bodily performativity" (Butler 2015, 9–11) is a visual occupation of city space, not in large public gatherings but in the fleeting moment of posing and posting a selfie. It is premised on the understanding that, of course, in breathing particulate matter, one is already more like an air filtration device than one would choose. In repeated digital images of cyborg appropriation through breath, bodies themselves appear as available for their disassembly into component parts and then their reassembly into the cyclings and phase shifts of the atmosphere, far beyond the sheath of the skin. Polluted air refashions respiratory organs into organic machines turned against the bodies they no longer refer to. City breathers, with the automatic, irrepressible need to inhale, appear in the hashtag campaign as individual elements connected by breath into dispersed, biotic mechanisms of a great, debilitating air-filtration device, which cycles the atmosphere through the urban populace.

Judith Butler, in her reading of Levinas, writes that the coherence of the individual subject is "undone, in the face of the other." To encounter another's face is not simply to be remade in the ethical demand in that face, but to witness the autonomy of selfhood, around which a liberal politics of individualism revolves, in its unraveling. In doing so, she seeks to re-describe autonomy as a matter of relationality, where the face of the other is an entry into peculiar *"modes of being dispossessed"* (2004, 19, italics in original). In the masked-face selfies, where individual faces are on display by being eclipsed, the ethical claim that shatters the autonomy of the subject into the interrelation of bodies implies a network of dispossessing conditions out of which other modes of relation might be borne, and out of which individual bodies might be shaped together.

Dispossession—condensed into hashtags and dispersed through particulate flows—proffers a politics in double negatives. Breathing in dense air, breathing into dense air, engenders collectivities and socialities that do not form for an agential assertion of rational self-awareness. Instead, this breathing is based on the vital debilitations that people cannot not take on through the autonomic process of breath. Bodily autonomy, for human vacuum cleaners, is a matrix for the forceful appropriation of the body at the contact of breath and atmosphere. And at the same time, that dispossession, shared through breathing-together, is the scaffold on an emergent body meteorologic, united only by what it does *not* want. The need to breathe, or rather, the impossibility of not breathing, makes haze not simply a question of the porousness of the body, but of the vital debilitations that make individual bodies each a part of a broader atmospheric becoming, dependent on its "thirst-quenching poison" (Li 2017).

However, as the increasing centrality of breath as a means of dispossession rather than of social unity suggests, the very presumption of the ubiquity of breathing belies the profoundly unequal distribution of breathing conditions. This distribution of life chances and death chances in air reconfigures images of political collectivity and the sites for the renegotiation of political demands into a geometry of processual atmospheric unfurlings and incorporations. Respiration is at once figured as a series of fractures in a coherent body politic and as the condition for other modes of political sociality. To render citizenship through the autonomic and nonconsensual biotechnical relation that breathing has become is to point to how particulate exposures might not just injure bodies but also make shared autonomic distress a tissue for some collective to come.

In reflections on the political potentials of air pollution in China, air is often approached as the trigger of a long-awaited mode of environmental protest, an efficient catalyst for democracy, a kind of liberalism with Chinese characteristics. Certain Chinese commentators also approach air pollution as an incubator for a new mode of political collective. Qiang Wang, of the Chinese Academy of Sciences, argues that "China is witnessing the beginnings of a civil society in which the Chinese people spontaneously defend their right to a healthy environment." It is the responsibility of the government, he offers, to "promote a civil society that stands up for the environment. The air of the people should be protected—by the people, for the people" (Wang 2013, 159). Grounded in the expectation of a long-delayed civil society finally coming into formation, such accounts search in patterns of haze for signs of a dormant political consciousness about to bloom into self-awareness. Euro-American commentators, mired in an array of post-Cold War habitus about the inevitability of liberal democracy, have searched for decades for signals of the fulfillment of this expectation. Its emergence would confirm liberal models of politics fixated on a metaphysics of resistance.

In counterpoint to the hope that air pollution might offer a respiratory trigger for pent-up liberal democratic impulses, these collectives of breathers also tempt with a vision of The People, *renmin*, central to Maoist mobilizational politics. Mun Young Cho, in her discussion of the urban poor in northeast China, considers the passing of The People, the privileged and empowered political collectivity and subject of Chinese socialism, into a "population" subject to technocratic intervention and management (2013, 18). Two visions of political collectivity contend, defining two poles of a typology of social forms structured by agency and its absence: one the muscular embodiment of ideologically corrected popular will, and one a patient on which power acts.

Cho argues, "The political platform of 'the people' is not entirely replaced by the numerical table of 'the population'" (2013). Her ethnography is not a story of transitions or subsumption, but one in which different ideas of collectivity contend with one another. The People returns, spectrally, to animate a political morality seemingly anachronistic to this moment in the disjointed time of the nation. Its return would be both a revival of waning demand for state responsibility and the resurrection, in the wreckage of the present, of a specific kind of muscular national subject.

Is the aggregate of "human vacuum cleaners" a return of The People? Is it a civil society rising in resistance to the state? To raise the question is to ask how a condition of exposure upsets a binary of agency and passivity. Breath, to those who wish not to become human vacuum cleaners, can generate collectives through atmospheric dispossession not only because it disrupts the liberal autonomy of the body and the self. It is the autonomic nature of the breath—the non-optional activity of the body as a material thing—that subtly displaces both the self-awareness and self-control of a civil society and the triumphal historical subjectivity of the socialist mass. For the shared endeavor of revolutionary life cannot directly be translated into shared breathing as a medium of common destiny.

Is there a way of asking for a collectivity that takes shape without the binary of the active and the passive? The autonomicity of breathing conditions a way of thinking through double negatives. Breathing is that which one cannot not do. To breathe is thus not an assertion of the body's agencies or the capacities of a collective, but rather the possibility of its dispossession through its own processes in the proper—wrong, violent—atmospheric conditions. If we return to deployments of the People as a necrotic mass that dies in order to live, or as vacuum cleaners made of human flesh, a populace of breathers emerges through a dispossession of capacities through exposure to the air. In this exposure, not only are individual breathing bodies assembled through the dense air, but bodies themselves could be dissembled into component parts and reassembled beyond the sheath of the skin. Air refashions respiratory organs into organic machines, elements of an atmospheric device that decenters and dispossesses the body of its agencies.

Exposure through breathing, neither active nor passive, might generate a vision of collectivity outside of a resurgent respiratory People or its disqualified mirror image, civil society. For breath hovers in a gray zone between action and automaticity, reflection and reflex. Like Chen's Smog Society, its very formation would take place in the set of complicated dispossessions and emergences of particulate events. The possibility of a becoming-atmospheric of the political thus lays in the dynamic tensions between action—liberal or socialist—and the inability to not act: between autonomy and the autonomic as grounds for a body meteorologic. It would require a reimagination of political life not as the assertion of agency but a field of intertwining and conflicting double negatives, riveted together through the smog that has not yet begun to break.

In 2019, while I was drafting this book, it felt to me for a moment that politics in China entered a state of suspension, rocked into sudden possibility from its Special Administrative Regions. In a Hong Kong that prided itself on civic peace and its postcolonial democracy, incredibly high costs of living and sustained worry over the future were engendering protests that occupied highways, skybridges, and public spaces. These demonstrations were largely organized by youth but included every sector of Hong Kong society.[1] They recalled the Umbrella Protests of 2014 when massive public occupations of squares, roads, and skybridges demanded the universal suffrage negotiated by British and Chinese authorities in the Hong Kong Basic Law as a condition of handover in 1997. In the summer of 2019, protests over a law that would have allowed extradition to China found protesters, clad in black, and donning gas masks and goggles, on the streets again. They were met with spectacular police brutality in the form of tear gas released over peacefully gathered crowds of students.

Tear gas might be "the exemplary medium of climate emergency." Its deployment is a mode of tactical microclimatic manipulation. Its appearance indexes political and ecological breakdown in "an authoritarian atmospherics, a militarized ecology" (Demos 2019). In the point-of-view videos that went viral across the Chinese world outside of the Chinese mainland and its tight control of information, the breakdown of the unspoken public contract that governed a liminal Hong Kong could be read in the thickening, caustic air. Popular movement hashtags depict the choreography of vapors as a concentrated state violence diffusing through and into the airways of gathered publics: trails of gas canisters flying into crowds, masses of peaceful demonstrators, keeping off the grass in the public parks, writhing apart to trace out the expanding toxic airspace engendered by the release of gas.

This police violence was a defilement of breathing life. It was the creation, in the open air, of intermittent spaces of air assault that turned the body against itself, much like the unbelievable air pollution that I had witnessed, breathed, and written of in Beijing. This gas, making the environment into a momentary no-man's air, ignited public support for the protesters across Hong Kong. It made the air, constantly a matter of comment and attention, suddenly forceful as the unspoken possession of a breathing public precisely in the moment of its weaponization. It made the state a meteorological condition. Gas gave the protest in 2014 its symbol, quickly circulated through social media as a logo and a meme: the yellow umbrella, which in the tropical island Special Administrative Region would be shelter from sun, wind, and rain, but also from a police apparatus that materialized as gas. In 2019, police gas attacks made the gas mask a palpable symbol of protest.

Sovereignty, following Aihwa Ong, is an experimental practice (2006). And gas, in this experimental assemblage, is more than gas. I remembered, fixed to social media, Deng Xiaoping's 1992 call, on the precipice of rapid market reform on the mainland, for Hong Kongs to bloom up and down the Chinese coast. China's experiment in Reform is premised on a poetics of urban reproducibility that takes Hong Kong as a model. If Hong Kong is, in gas, to become Chinese under atmospheric duress, then, China, too, has for decades been becoming Hong Kong. China and Hong Kong do not relate in stories of integrations or secessions, but rather in play and ambiguity of experiment and example in which each refracts the other. The trails of gas manifest more than the savage exercise of state violence, as if Hong Kong's civil authority were only a proxy and client of its Beijing benefactors. In addition, another kind of experiment came to the fore, as if authority were feeling its way through the dense atmospheres of the subtropical Chinese place as a prelude to testing out a vision of China as condition of being-gassed.

To register the fact of exposure is to feel an existential tethering to something beyond the body and its tightly demarcated socius. In gas, spilling out onto public grounds, this tethering is folded into a political tactic for the sudden configuration of an environment into danger, an atmospheric desert in the stifling tropical heat: "crowd control." Tear gas, in these terms, is a technology for weather-worlding, where state terror metamorphoses into techniques for engineering the very meteorological conditions for existence, to overturn the relation between life and breath. The public-in-formation shifted quickly in this moment. The subtropical air became evident as the always-present ground for the movement as well as its potential undoing. Its quick appropriation, through toxic gas, made it evident as a volume that, for political purposes, would be uninhabitable before it would be public in the form imagined by protesters.

Vaporizations: "not an obliteration, but a shift in phase, a distribution into concentrations" (Choy and Zee 2015, 217). In my last trip to Hong Kong, after one protest and before the next, I felt this Chinese air as a tissue that held my body to its hazy histories. Humid air grips, folding the body into its density. It would not let me leave. I remembered the stifling summer wet of my grandparents' ancestral home in Ningbo, China, which we visited together just before my fieldwork and several years before their passing. The village faced the infinite Pacific. We turned away from the ocean, looking over new bridges toward the city, rising as one of Deng's Hong Kongs, with its lights and promise. With aunts and uncles and cousins that my grandparents always remembered but I had just discovered, the subtropical night pressed. The weight of China, of Chinese-ness, gripped me as thick heat.

On Instagram in 2019, I watched the Hong Kong police fire gas canisters from behind walls of riot shields, and gas-masked protestors contain and redirect their toxic release with traffic cones. I felt myself holding my breath, suddenly spirited away to my grandmother's village. The involuntary tightness in my chest was an arrest into a sympathy I could not extricate myself from: a forcible breathing with others. "Breathing together, sharing a destiny," when I read it later, felt like a shock, and a theory and threat of Chinese-ness in deadly air. I felt this strange interpellation into an ethnic respiration, for friends in and from Hong Kong to be sure. But I also felt, in retrospect, my entire kinship chart, real and fictive, its spokes stretching into China and a diaspora charted in vulnerability to the air. In images of gassed protestors, I saw a spectrum of my family's faces, of rediscovered branches of a clipped tree. All wait for the relief of the evening downpour, of the plum rains of May.

In this book, I do not write much of democracy. For decades, foreign commentators have done so, with their arsenal of human rights complaints and their readiness to read an incipient America in Chinese political collectivities. Whatever form of collectivity may emerge in China—is already emerging—is reconfiguring out of its variegated political and environmental geography. And it will emerge from the historical conditions of the country's shapeshifting socialism, wrought as a profoundly generative and experimental political formation. And in any case, America these days looks nothing like that promised by liberal teleology. In 2014 and 2019, the comparison of Hong Kong to the Beijing of 1989 was all too evident, circulating wildly in public discourse. This was not merely for the state violence that could not but appear as the echo, decades later, of other great gatherings—which were now tightly policed, if not altogether banned in the mainland.[2] Not so much backslide, but the announcement of something to come through the power of a reminder that could not have been lost on any party involved.

But in 2014, the well-organized encampments that for weeks replaced the city's downtown thoroughfares took shape with and in the wakes of gas. The sudden need to find shelter in an air rendered toxic had also, for a moment, created an ersatz collective held together by their subjection to the state through their vulnerability to its airs. In 2019, Hong Kong had again exploded in protest and retaliatory, world-shaping gas. The police, against all manufacturer warnings not to use tear gas in enclosed spaces (Feigenbaum 2017), shot canisters into subway stations and trains.

Protest tactics had shifted from attempts at sheltering against or escaping from tear gas to a choreography of police atmospheres.[3] Protestors, borrowing from Bruce Lee (see Lee 2020), encouraged each other to be like water: flowing, hardening, both passive and powerful. In their imagination of protest subjectivity through the formal processualities of phase shift, they also learned to manipulate the flows of noxious gas that defiled open airspace. The umbrellas that sheltered protesters in 2014 could be raised *en masse* to throw up tactical atmospheric architectures, protecting protestors in tortoise shells of charismatic yellow, a phalanx formation against bad air. Gas masked demonstrators, becoming like water, learned to make gas flow and pool, provisionally choreographing the atmospheres that were defiled to injure them. Black-clad protestors met tear gas canisters with municipal traffic cones that would contain their toxic payload until it sputtered out of the cones' open tips like science fair volcanos. Turtles swim in this unfree air, the noxious ocean their sky has become. Protest and policing had become an experimental atmospherics.[4]

A hemisphere away, I lose days to clicking through the warren of hashtags that document the event in real time in English and Chinese.[5] I guess at the nuances in their defiant Cantonese, rife with characters that do not exist in the Mandarin of my family and the *lingua franca* of my fieldwork. I lose more time retracing my steps, swiping to refresh the raucous feeds of images and slogans. One image from 2014 has stayed with me the most, though I can no longer find it. It "cannot be described, nor simply shown but gestured toward" (Stevenson 2020, 9): a couplet written vertically in black on two white banners echoing, in form, the blessings that people post in color aside their doors to welcome the New Year, but in this case in the stark black and white of a funeral procession—a mortuary celebration. The tight formality of the phrasings drew from a visual language with gravitas, a mourning in the aftermath of the violence that had impossibly crossed a terrible, symbolic threshold.

I have lost the image, but it is so clear in my memory. It is of an old man, serene under an overpass, surrounded by younger protesters. He is holding the hand-painted banners. The couplet is from a poem by Bai Juyi, a Tang Dynasty poet. It describes grass on a devastated, scorched plain. "Wildfires cannot burn it away / With the blowing of the spring wind it grows again" (*yehuo shao bu jing, chunfeng chui you sheng*). In the bluster and then the calm of the demonstrations, demonstration itself emerges as a weedy fortitude, a scrubby resurgence, a tenacity in small things. It is the action of a collectivity that is not yet a subject, drawn together through sympathy, protest, and the vulnerability to policed atmospheres.[6] A meadow born of conflagration. In its poetic vision, social movement

powers this collective not by the triumphant rise of a self-realizing movement against power but in the quiet endurance of wind, fire, and seed,
architects of a crushed world.

In the *Analects*, grass and wind are pedagogic elements in the illustra-
tion of a theory of righteous government. "If you desire goodness," Confucius exhorts, "then the common people will be good. The Virtue of the
gentleman is like the wind, and the Virtue of a petty person is like the
grass—when the wind moves over the grass, the grass is sure to bend."
Later, the sage teaches rulers, "You are the wind, and the people below are
the grass" (2003, 134), a great virtue bending lesser ones by its righteous
example. Two depictions of the grasses and the winds: two depictions of
political cause and effect. One where the wind cows the grass by its right
force, the other where grass and wind rise together as two elements in a
perpetual and perpetually changing universe. So many worlds are drawn
together in this relation of great and small things, out of whose relation
the rise and fall of states and rulers can be imaged.

Across the Pacific, I learn, in long walks through the redwoods that fire
is ecological. It renews as it destroys, Andrew Mathews shows me, scanning the ground near the road cut for the cones of knobcone pines, which
need heat to open. In the long knobcone winter of fire suppression (Pyne
2015), they have been overtaken by other trees. Knobcone pine, like the
grasses that return with the wind to a scorched earth, waits for its burn.
They grow because of it, not despite it. Destruction—the fire or its denial,
the gas that begets the umbrella and its movement—incites form. This
political ecology of poetry and protest is one of renewals and intensities.
Tear gas disperses a crowd only to gather another. In the invocation of
the poet's scene of grass, fire, and wind, renewal comes with all the inevitability of the cycles of a dynamic universe, as natural as the winds' rise.

5 City of Chambers

GRADIENT SPATIALITY AND A POETICS OF DUST

CANNED AIR

In early 2013, as Beijing was enshrouded in record air pollution, recycling entrepreneur and billionaire Chen Guangbiao took to the streets. Dressed as a large orange aluminum can, he handed out smaller orange and green cans emblazoned with his logo-fied face to curious passersby. The cans were empty. Or rather, they were filled with a consequential nothing: each contained a volume of compressed, fresh air, devoid of the fine, dense particulate matter that hung in Beijing's soupy atmosphere. The air came in multiple "flavors," canned at the atmospheric source: pristine Tibet, whose distant, upwind airs remained purportedly untouched by the adventure of Reform and "socialism with Chinese characteristics"; revolutionary Yan'an, the hallowed red capital of the 1930s and 1940s, purified in the singe of Revolution; and postindustrial Taiwan, an extant model of a clean Chinese alter future. These flavors charted a geotemporal trajectory that linked the past and possible futures of multiple Chinas (P. Liu 2015) differentiated into experiences of breath, packaged and shipped to the capital. The jovial entrepreneur distributed two hundred thirty thousand of the cans for free. He sold another twelve million at the hefty price of 5 yuan a piece, just less than 1 US dollar.[1]

The cans were released into an atmosphere of suspicion. Canned freshness and the respiratory vulnerabilities it suggested recast distressed respiratory systems as a key embodiment of the failing promises of the invincible juggernaut of Reform and Opening, choking on their fallout of lung-penetrating particulate matters. Late socialism, for China, was a terrain of faltering atmospheric and respiratory care. Indeed, Chen states that the canned air is in part a contemporary reactivation of investment in the integrity of a socialist body politic, one left exposed to the wild of a modern air condition. Half performance art, half marketing ploy, but given entirely as a service to a nation of beleaguered breathers, the cans ironically replicated, in miniature scale, an ongoing splintering of the country's airspaces into conditions of differential breathability.

Each can might be read as a small moment of what Peter Sloterdijk calls an "air conditioning," the "disconnect[ion of] a defined volume of space from the surrounding air" (2009, 20). The can's hermetic airspace is a promise of the disconnection of a pulmonary system from a generalized condition of exposure, achieved through the technical and commodity intervention of the single-serving can of nothing. Their emptiness is at once a promise of enclosure against dense, dangerous air and a deliverance to a respiratory elsewhere, a portal to an alter-China in a gradient of better air conditions. Their supply chain of mobile Chinese airs converges in Beijing in an artificial national airstream, a strange double of the dust-storm routes that cause a major problem of sudden staggering increases in particulate counts.

The proffering of a breaths-long respite from urban air, through the patriotic spectacle of cans on sale, emphasized only what had become, by the winter of 2011, unbearably and unbreathably obvious: in this city entombed in the startling density of its atmospheres, breathing had become a modern danger, tying urban bodies and environments, life and its erosion, together in a toxic knot. This is a medical and meteorological irony of Reform, substantiated in the exposed body-in-air as a register of a national condition of stunted life: as the standard of living rises in the hypnotic mirage effect of official statistics (X. Liu 2012), day-to-day life shades into conditions of unlivability, a search for pulmonary shelter. Commodified air ironically married markets and biopolitical nationalisms, so that, for a price, China's urban everyman could breathe, at least for a moment, like officials in government buildings, whose corridors and offices were rumored to be lined with expensive air purifiers that remitted respiratory privilege as a perquisite of political position (Jacobs 2011, A4).

Certainly, commodified air enacts a gross primitive accumulation of air, what Elias Canetti calls "the last common property" (1979, 13).[2] But the cans—and their tantalizing promise of respiratory shelter in a handheld serving size—must also be taken as an incitement to consider a city and its compartmentalized atmosphere as an emergent volumetric formation of late socialism and its perpetual weather trouble. As atmopolitical commentary turned surprisingly good business, the cans dramatize a search for breathing space in conditions of meteorological insecurity turned persistent public health crisis. And they speak to airspace not simply as a meteorological background, but as a domain of technical experiments and interventions: its variable densities, the allure of its decomposition into hermetic volumes, with its poetics of volumes and bodies all sites of a technical, political, and bodily negotiation subject to provisional modification. The cans are, in this way, a template for an emergent mode of atmospheric strategy and urban morphology in an everyday rendered as a low-grade dust storm.

The cans resist being dismissed as only the straightforward monetization of the air and the splintering of atmosphere and its breathing publics into a pay-to-play marketplace of breather-consumers or simply of a rapacious market finding its most cynical opportunity in the enclosure of air, a last, broken commons.[3] They demand an attention to enactment of city space as the promise of a hermetic *airspace* as a quotidian technical and embodied tactic of finding and making air shelter. I am after what the cans, for all their apparent absurdity and perhaps their actual cynicism, offer as an entry point into the remaking of a voluminous cityscape.[4]

As an atmospheric technology, they circulate through airspaces already in the process of technical and political modification in a chorus of bodies, humming machineries, and architectural openings and enclosures. They anchor, in their circulation, a mode of inquiry into the urban in the haze of a city identified as much with its sky as with its buildings: a city registered through its recomposition into aspirational breathing spaces (N. Chen 2003), from a sea of masked faces to the tree-walled city itself, a much-too-open airspace. Urban life and form appear, then, as an interplay of merging and splitting volumes, each a moment in a gradient of dust and its densities.

———

In the seminal essay "Walking in the City," Michel de Certeau issues an injunction to delve into city life for "a theory of everyday practices, of lived space" (1984, 96). This theory would be composed through the myriad spatial practices through which cities are made and remade by their inhabitants. The ambulatory view, embodied and at street level, recuperates the quotidian as a site of spatial and knowledge production, against the cartographic abstractions of the urban planner's vertical view. Walking, he famously argues, is a mode of capture and resignification of space, requiring a foregoing of vision from above (see Kaplan 2018) into the ordinary geometries through which space is lived as experiments and contestations. This is a forceful everyday act of resituating the city through its inhabitants rather than the specific abstraction of the aerial gaze, exemplified in the view of New York City from the observation deck on the eighty-sixth floor of the Empire State Building.

This chapter is after a poetics of particulate matter: a substantial and volumetric vernacular for attempts at reengineering airspaces and their interfaces with breathing bodies. It poses de Certeau's question in a different way, in the episodic particulate density of China's capital: what if walking in the city was also breathing in the city? If it was not merely an appropriation of urban spaces but also an appropriation of bodies *by* these spaces in the relentless breach of inhalation? De Certeau's question unfolds into dimensions, compelling attention to an experimental volumetrics of urban life. Architectural inhabitation appears as a practice of "creating the dimensions in which humans can be contained" (Sloterdijk 2011, 28), an architecture of airspacings in the gradient density of the air.

I trace an array of techniques, installations, and "science art worldings" (Haraway 2016, 71) to witness the proliferation of airspaces in the city. Airspac*ing* and its poetics of dust are technical interventions into atmosphere not as an abstract surround, but as a voluminous, three-dimensional materiality that may be partitioned, ballooned, or manipulated as a particulate gradient. We begin with Chen's canned air, and extend the previous chapter's questions of meteorological embodiment into the emergent technical atmospherics of Beijing's famously dense air. This orients us ethnographically toward strategies and practices that seek to filter, capture, condense, and otherwise act into aerosol atmospheres to ford what the central government promises is only a passing phase in a guaranteed meteorological transition.

Chen Guangbiao's cans of air help us to think the city through airspacings: atmospheric technologies through which a poetics of shifting volumes and gradients operates. They help spur inquiry into not only the scale of air pollution crisis, but how changed air conditions proliferate strategies and technologies for inhabiting and remaking urban life and space in explicit atmospheric attunements to the properties of fine particulate matter. The techno-poetic practices that I track require "looking beyond the metaphors that shape our experience" of a given atmospheric condition and attending "to the material realities of our current environmental crises" (Byrnes 2018, 20); airspaces are not simply a metaphor of separation, but situated practices of intervention into the properties of atmospheres and the manifold modes of existence that they afford or preempt.

The cans, as a techno-poetic intervention into particulate airscapes, alert attention to a theory of urban life as a choreography of dusts and particulate flows and densities. The airspaces that they seek to unfold out of Beijing's atmosphere in turn draw attention toward the experimental practices, logics, and modes of atmospheric intervention, which attend to life in the air as a mechanics of particulate physics. Together, they disallow the identification of the city as such with either its architecture or the quotidian heroism of individual breathing body-subjects. Our urban attention instead wanders to a conception of urban air that is already more or less ubiquitous in practices of airspacing in Beijing. Dusts, soot, and particulate matter have revealed airs, across a variety of domains, as a massively dispersed, and provisionally manipulable semi-solid.

Manipulations of airspace enact urban space as a question of atmospheric morphologies and gradients of particulate density. The voluminous city of enclosures, shelters, leakages, and densities that takes shape in the dust is an urban poetics appropriate to life in the suspenseful temporality of aerosol becoming. I am after an experience of breathing in a city that, in the changing moods of its sky, folds, balloons, and vacuums into a morphology of nesting volumes, never quite well-insulated enough from particles that still seep in. This city should be located not in the spectacularity of its arriv-iste starchitecture, but in the manipulations of aerosols, particulates in/and air. Urban meteorology is to be located not in the city's beleaguered airspace, but in practices, infrastructures, and events of airspacing: the manifold technocultural conjunctures, art and design practices, and, indeed, practices of everyday life that make, of a city suffocating in haze, a city of chambers.

As buildings fade into the contracted visibility of the urban pall, a dif-ferent city actualizes with haze. It is composed in the interplay of breath-ing spaces and the variable densities of particulate matter, where everyday life has become coterminous with practices and techniques of airspacing. In recent years, anthropologists and geographers have asked after atmo-sphere, with its turbulences, freights, and morphological vibrancy, for a poetics of air. One way that air comes to matter is that it inspires "a mode of attentiveness to deeply affective and personal resonance of airy matters" (Engelmann 2015, 432). Interest has abounded in the protean forms that air shapes and that shape air, in various degrees of durability and mul-tiple images of flow: balloons (McCormack 2018) and aerostats, (Shapiro, Zakariya, and Roberts 2017, Kaplan 2018, Coen 2014) whose floatings trace out the mercies of air currents. Nerea Calvillo's data art project, "In the Air" abounds with meshworks that visualize atmospheric data as fluctuating nets that hang over and slide through city buildings. These overlapping grids forcefully recontextualize the earthbound architectural space in ref-erence to "a 'diffuse façade,' a massive indicator of the air's components through a changing cloud, blurring architecture with the atmosphere it has invaded and mediating the activity of the participants it envelops."[5]

If everyday attunements to air emphasize that it is not a passive meteo-rological background, but rather a dense and dynamic composite substance with substantial and spatial properties, urban space might be reimagined in both the centering and technicalization of breath. Airspacings might be understood as part of a repertoire of minor practices for dwelling in con-temporary air: in other words, a mode of urban living defined in the passage between airspaces as part of an everyday practice that attunes to atmosphere as a scheme of possible risks, accepting some damages as unavoidable.

To capture this, I home in on the intermittent and semipermanent ways in which attempts at living, breathing, and building Beijing's meteorologi-cal specificity have revealed urban air as manipulable airspace, a poetics of aerosols and atmospheric densities. I explore practices that figure the capital as a shifty three-dimensional and real-time architectonics of particulate strata and nested volumes. Architectures, filtration apparatuses, and endangered pulmonary systems recur in new urban planning rationales as well as in emergent architectural vernaculars. I learn from Judith Farquhar and Zhang Qicheng, who offer a masterful account of Beijing through the array of vital practices for nurturing life, *yangsheng* (2012). These practices construct urban life as a quotidian vital public, giving a slant view on the grand pretentions of the state and its spectacular designs and concomitant anxieties. These vital practices are at once sustaining and defensive. They are oriented both to the cultivation of a fuller life and also to preparing bodies to undergo less damage.

Their account of techniques of the body also revises bodies as unwitting atmospheric technologies as breathing and different air atmospheric devices converge uncannily in this chapter's focus on a reimagination of bodies as technical environmental systems and sensors (cf Myers 2015). Like the previous chapter, with its focus on unexpected breathing publics, I seek to offer an image of the city through aerosol predicaments. Here, we approach the city not as a patchwork of cartographic spaces, but as a complex process of interventions in air conditioning: the technical manipulation of meteorological dynamics and particulate densities, on the one hand, and the attempt at insulating hermetic breathing spaces out of an increasingly untenable "general" air, as when lips seal over an empty can rather than the air around it.

In the decades before the Chinese government had promised the resolution of the particulate problem, dense air gestated new spatial and substantial formations into and against aerosol particulates. Practices of airspacing, especially through practices of technically mediated disconnection and conditioning, offer a way in to the productivity of air pollution as a material and political phenomenon and thus to attend to the near future as something besides what central officials assure is only a period of waiting until a solution is achieved. Living in and as a part of a particulate suspension is thus to also be living in suspended time. The remediation of pollution and general breath will, in the best-case scenarios projected by Beijing's municipal government, require decades, an entire generation of diminished life.

The attempt to hollow out breathing spaces from dense air must be understood as part of a broader strategy for making sense of life in *this* air.

It draws our attention to the dislocation of breathing bodies beyond the skin and of the lung cavity as an open-ended contact of body and environment. For the exceedingly well-heeled, the desire to breathe freely—and especially the intergenerational desire for healthy children—culminates in fantasies of escape, an extrication and installation of bodies in other airs. The migrations of birds, after all, index the viability of ruptured ecologies (Guarasci 2015). Beyond fantasies of flight, another city is taking shape.

The chapter moves between two broad atmospheric thematics. First, it orients toward practices of airspacing in a diverse array of domains that move between middle-class dwelling, the experimental conditions of air-quality experiments, and architectural forms that employ air as a building material. It asks how China's meteorological contemporary—registered in the aerosol entanglements of land, air, economy, and bodies through breath—demands an attention to space as a volumetric practice of generating technical airspaces that seek to approximate "normal" breathing conditions.

And second, I explore two art projects addressed to Beijing's air as a massively distributed solid rather than a passive emptiness: a pollution-catching tower and a tedious repeated work of vacuuming the sky. These two projects, respectively the invention of a Dutch art and design firm and the viral social media object of a Chinese performance artist, both approach the city's air through intake devices that precipitate dispersed aerosols into particulate solids, demanding an attention to air as a gradient composition of dusts. Through these projects, we can consider the lived form of the Chinese city through the spatial and substantial intervention of idiosyncratic atmospheric architectures. Breathing in the city reveals it as an architecture of nested volumes rather than gleaming surfaces, a city of episodic interiors scooped out of the dangerous sky.

CITY OF WALLS, CITY OF CHAMBERS

My friend Xia Jie visits our probably illegal apartment in the *hutong* alleyways that in 2011, the main period of my fieldwork in Beijing, had survived mass clearings for the 2008 Olympics and preceded the municipal crackdowns that would demolish and brick over off-code constructions a few years later (Palmer 2017). It is my first winter in Beijing, and probably my first *real* winter ever, and I still insist on carefully ignoring air pollution and the ubiquitous smell of burning coal in the city sifting through the cracks under our prefab plastic door. The only ones who seem to notice air pollution are the loud Americans at the bars we sometimes go to, who announce air-quality readings in units still unintelligible to me.[6] The effect is one of scale, an expatriate complaint figured as an education in feeling numbers, coordinating them with feelings that I meticulously unsense as part of my life in the city (Miéville 2009).

Many of my conversations with Xia Jie circle around my general ineptness in the harsh north Chinese winter, having spent my entire life in California. Weather is a shock to me and an ethnographic fixation. On a particularly gray day, she bursts through the door with an armful of materials and accoutrements for winterizing the drafty illegal construction: hot water bottles and heating disks that charge by USB as a radiant portable heat source. The smell of hot plastic. Adhesive sheets for double- and triple-layering the thin glass windows, and cheap heaters she's ordered on Taobao on one-day delivery. But first, she insists on sealing the crevices around doors and windows, laying damp towels across jambs. This is to separate not only the apartment from the outside but also the rooms from each other. Otherwise, the apartment and its shoddy structure would be a short chain of airspaces still perilously abutting one another through minuscule cracks that could never be well enough sealed.

I get the distinct feeling of walking around the apartment as a moving deeper and deeper into a cavern, like descending through the colors of a state air-quality scale, from off-scale purple to still-upsetting yellow. Before she leaves, she puts on a 3M painter's mask. At the time, it feels excessive and a bit alarmist, and I can't remember how my body felt, as if I were carefully unfeeling it so as not to too closely align with other Americans checking the US embassy's Twitter feed with a grim schadenfreude. A refusal of the *co-* in that commiseration: "China!" After all, the last time I was in Beijing, we waited in a mall, with masks hanging around our necks, just for a while before going out again.

––––––––

The logic and engineering of self-contained airspaces can be traced at many scales in a Beijing of airy enclosures and harbors. Peter Sloterdijk calls this entry of air and breathing into the field of intervention "air design," the "technological response to the phenomenological insight that a human being-in-the-world is always and without exception present as a modification of 'being-in-the-air'" (2009, 93). Air design, which Sloterdijk associates with the advent of modernity as a peculiar style of atmospheric attunement, appears in his work first in its special relation to engineering death through weaponized atmospheres like chemical warfare or gas chambers and then in the philosophical-technical realm of engineering life supports for spaceships that have to carry and replicate Earth inside of them (Sloterdijk 2016, 315–22; see also Battaglia, Valentine, and Olson 2015).

Air design in this chapter offers instead a way into exploring the attempts at realizing pockets of planetary habitability in the midst of changes that may not find resolution. They spin out in technical practices that begin from forced attunement to atmosphere and breath that ultimately offer a view on urban habitation as the manipulation of aerosol morphology. What provocative skyline might be revealed when the city's monumental hyperbuildings dissolve out of view (Ong 2011, Koolhaas 2004)? And when the hum of air filters and the blooming of face masks carry the hope of insulating so many breathing spaces in the Chinese capital? And when a city of spectacular facades empties, mercifully, into containers of purified air?

A boundary between interiority and the exteriority is becoming not only physically but conceptually permeable, breached by particulates that sneak through cracks and penetrate walls. Lisa Rofel closely ties shifts in the texture of socialist life in China with the changing meanings of inside and outside, a "perverse cultural construction of space embedded in implicit knowledge (1999, 64). Inside/outside peels away from the hard mapping of structuralist binaries, opened into a state of flux. Debates over urban air quality make this reconfiguration of hard interior and exterior distinctions literal, as concerns over outdoor pollution tip into anxieties over indoor air, as in Xia Jie's insistent clotting of the cracks between doors and sills. This suggests that air pollution, as a quotidian feature of city life and an emergent political question, must be approached not simply through macropolicy projections over long-term air quality amelioration but through micropractices that render the air as a cascade of fluctuating substantial and spatial configurations.

As aerosols hang heavily on windless days, even the "outside" can be understood increasingly as the inside of an ever-larger containment, where the mountain ranges surrounding the city block air-flushing winds and hold Beijing in a basin for the catchment of pollutants. Broad boulevards are redesigned as outdoor ventilation corridors (*Xinhua* 2016) by removing obstacles to airflow. They aim to transform city buildings into channels to flush out particulate haze, inducing the formation of powerful wind currents in urban street canyons that planners seek to avoid in other places. The city's architecture is to be realized into a network of wind tunnels, opened out at the level of the city itself through the removal of obstructions.

Wintertime thermal inversion, wherein layers of cold air form a "cap" on the city's basin, tends to create air as its own container, sealing the city against the winds that might flush out collected particulate matter (Feng et al. 2014, Chan et al. 2005). Air pollution episodes are linked to the vertical, meteorological architecture of the city, where thermal inversion events effectively enclose an airspace, an atmospheric trap that holds particulate and other pollution low and still. "Stagnant air masses" settle over the city, creating conditions in which "solar radiation could not pass through the haze" (Ji et al. 2014, 548, 549). In these conditions, the city's atmosphere is no longer the outside against which insides are distinguished, but rather, it becomes a first great interior out of which others are carved. It is a cavernous heft to be acted on directly and *in situ* in a replumbing of civic airstreams. Exteriors become massive interiors, and the city, a mountain-walled, air-locked pocket of particulates, suspending and settling in the circulation of inside winds.

Within such an enclosure, other enclosures balloon. Beijing has long been a city of walls that transect urban space in a serial, fractal elaboration: the Confucian compound and then the *danwei*, the socialist work unit, each generated a cellular urban form of walls and compounds (Bray 2005, Lu 2011). The *danwei* system reorganized social and political life, government and administration, social services, and kinship around production and work. In its most typical form, the *danwei* was a miniature and self-enclosed city, centered around a factory. As David Bray shows, the *danwei* spatialized socialist planning and the ideals of productive activity as a format of new social relations, in the "basic cellular structure" (2005, 120) of the new Chinese city. It repurposed the ubiquitous walls of cities like Beijing as "a multiform architectural technology" (2005, 19) for the experimental form of socialist space and power.

Complex social relations and spatial form echo each other in the built environments of imperial and socialist planning. But more, the city itself could be continually decomposed into smaller and smaller units, for a kind of cellular-fractal urban formation built out of *danwei* that more or less functioned as distinct subcities, each marked off with walls. The walled compound has continued as the basic cellular-spatial unit of the Reform-era gated housing community (*fenbishi xiaoqu*), "a new pervasive spatial form that organizes life in cities" (A. Zhang 2020, 93; see L. Zhang 2010). The conditioning of airspaces expands this subdivision of the urban plane into three dimensions, cutting not only up from the ground but also across the sky: a city of walls cutting across space foreshadows a city of chambers bubbling as volumes out into air.

These chambers, like the *danwei* work units that they evoke in three dimensions, are not rejections of the city-as-airspace, like postmodern architectures that reject their own urban context, "for [they] do not wish to be a part of the city but rather its equivalent and replacement or substitute" (Jameson 1992, 40.). Rather, in their attempt at generating airspace out of a general air, they rework the distinction between the general and the specific into serial and variegated iterations of a singular aerial form, disclosing the city not as a unity but as an aspirational, desperate fracturing into discrete and yet-too-interconnected airspaces, an urban morphology of adjacent and nesting volumes. This dual movement of escaping city airspace by replicating its volumetric form means that we should not approach a city of chambers as a proliferation of escapes but as a burst of atmospheric shelters, each of which can be analogized to every other at different scales. Their agglomeration, even while striving for a complete atmospheric disconnection from one another, is the very condition for an urban form that continues to be realized in a congregation of cellular spatial units, room spaces (Kelly and Lezaun 2018) that replicate and iterate the mountain-walled, thermal-capped city itself on days of stifling particulate pollution.

In early 2014, the Chinese Academy of Sciences submitted plans to build a massive indoor smog-chamber facility in the city's Huairou District to generate and simulate air pollution under controlled conditions for study at the whopping cost of 500 million renminbi, US$81.4 million. While certainly not the first smog chamber in the world, or in China (Wang et al. 2014), or even in Beijing (Wu et al. 2007), the project, announced as the largest in the world, was notable for both its scale and for its political exigency, replicating the city's air pollution in the experimental containment of its massive chamber. When built, the chamber will enable the study of atmospheric photochemical reactions in a variety of air conditions, through precise mechanisms for controlling the temperature, chemical, and particulate composition of the experimental atmosphere, which is then irradiated in order to establish reaction models (Wu 2007, 250).

The chamber must be understood as the technical implementation of an atmospheric logic that separates airspace in order to replicate it in a different iteration. It is perhaps a distant descendant of Boyle's famous air pump (Schaffer and Shapin 2017) and its intertwining of early scientific veridiction with the manipulation of airs (see Rogaski 2019). But it should also, crucially, be apprehended as a mechanism through which a clear distinction between the possibility of delimiting an absolute distinction between insides and outsides erodes. The chamber works experimentally by enacting the conceptual interchangeability of interior and exterior atmospheres, an indoor modeling of the erstwhile open. The massiveness of the smog-chamber facility in Beijing is certainly a move in a politics of spectacle and scale, an extension of a political aesthetics in contemporary China, where distinction is often offered in superlatives: biggest, highest, fastest. But it also aims at the technical and epistemological obliteration of scale by reconfiguring the chamber's interior and the erstwhile exterior that it seeks to replicate as an experimental system.

Larger chambers generate atmospheres large enough to approximate the outside; an inside large enough to be an outside, its size is part of the methodological and epistemic exigencies of smog-chamber research. Size is one way of reducing "wall effect," the disparity between "real" air and its experimental approximation, which can be linked to the fact of its containment, in the settlement of materials or reactivity of chamber walls. The technical reduction of wall effect—through ever-larger chambers and through the use of materials and coatings—names the aspiration to obliterate the distinction between interior and exterior airs, or, better, to render each a permutation of the atmospheric as a formal interiority. Wang Gengchen, a researcher at the Chinese Academy of Science's Institute of

Atmospheric Physics notes, "The larger the smog chamber, the better its simulation quality" (Fang 2014), despite the other inconveniences of size (Xinjing Bao 2014)—like Borges's fantastic map, coinciding point for point with the space it represents (1981).[7] Two atmospheres, then—not opposed as one contained and another open, or as original and simulation, but linked as iterations of one another: the city's atmosphere as the inside of a great smog chamber, a giant can of air.

THE DOME

As air pollution wears on as a fact of life and breath in Beijing, conditioned airspaces become an increasingly important architectural form, perhaps someday to change the face of the city as much as the more spectacular boom in landmark buildings. Air design has become an everyday practice, especially in transforming domestic spaces into protected airspaces. The architecture of air in Beijing seeks a final autonomy of airs from one another through a technological extrication and especially a fracture from the "great air" (*daqi*) of the city. In a city of single children, the middle-class investment in the bodies and prospects of children (Anagnost 2004) has made schools and homes significant sites of intervention, within which a new built environment of the air is hollowed out.

In 2013, the International School of Beijing, a private school for the children of expatriates and wealthy Chinese, made headlines in the international presses when, in response to parent complaints and an exodus of expat families from the city, it erected a huge pressurized dome over the school's outdoor playground so that children could play in an artificially generated "outside" on polluted days. Stories of such domes circulated in the international media as smug proof of the severity of the city's air problem, claiming that in the domed playground bubble "childhood" itself might be delivered from the mortal danger of breathing (Wong 2013).

On days where the official Air Quality Index (AQI) spurs government warnings to reduce activity, and thus the need to breathe, such air-supported structures allowed for a normalized operation of the body through the maintenance of a secure, sanitized atmosphere. Sourced from subsidiaries of companies in the United States, where they are used for the enclosure of stadiums, these massive synthetic skins use air itself as a construction material: they hold their shape and preserve air quality through the constant operation of machinery for manipulating air's form and composition.[8] After the air passes through filters and is expelled through the pressurizing fans that continually generate the dome's air structure, interior air-quality sensors register negligible levels of PM2.5 even when just beyond the enveloping walls of the dome, numbers are so high as to be off the charts.

The dome takes on the shape of manipulated air as its primary structural principle. It indicates a doubling of empty space, *emptied* space, into the atmospheric pillars for protected childhood. The dome also burst into prominence as an emergent ur-metaphor for Chinese air more generally in 2015 when a documentary on air pollution in the country quickly passed 150 million views before being summarily censored. The journalist Chai Jing's film *Under the Dome*, a TED talk–style lecture turned viral media event (Yang 2016), offers a dramatic account of air pollution as it has transformed the textures of life and death in the country. It opens with an image of air-endangered childhood, the outline of the narrator's fetus in ultrasound, stricken with a tumor *in utero* that required immediate postnatal surgery. The image of the city as a dome of suffocating particulate matter poignantly recalls that of an incubator in the neonatal intensive care unit. The enclosure of outdoor areas against outdoor air occurs at the confluence of disparate elements and processes: biopolitical investments in children conditioned by Chinese population policy and lingering even after the relaxation of the so-called One Child Policy; hopes and ambivalence in economic development; and the centrality of breath as a discipline of the body and of air in Chinese conceptions of health and illness.

While for now the air domes remain mostly in private schools one-upping each other for enrollment of the urban wealthy, their logic of enclosure and filtration is becoming much more ubiquitous, diffusing out from a concern of the anxious rich to a broader middle-class problem. The ubiquity of the air filter as a technical appendix for respiratory shelter has given way to a mode of technologically mediated human existence that Matthew Kohrman calls "filtered life" (2021). Consumer air filters churn away in the paradisiacal compounds of the rising middle classes (L. Zhang 2010), and glitzy shopping malls announce their HVAC setups as enticements to shop.

Managing the air is becoming a technical and architectural practice through which the city spaces are changing. Whereas air domes work through the pressurization of air as a structural component of enclosure, filtration technologies allow the conditioning of any room into such an airspace. Filter technologies begin from the problem of indoor air rather than air quality writ large, especially the suspicion borne out by pocket air-quality-measurement devices that even the indoor air is not safe. In this, they operate at the much more concrete interface of body and highly localized breathing environment and aim to create intermittent pockets of cleaned air. Rather than clearing the entire atmosphere, filters allow for the multiplication of smaller breathing spaces carved out one room at a time, to create a breathing space that surrounds the body as a microsphere hugging close to the mouth and nose.

Rooms are transformed into containers through the fan-driven forcing of air through filters. But bodies are unsafe even in the deep interiors of unfiltered domestic space. Where the air threatens, this human body itself takes on the characteristics of a compromised airspace: it extends past the porous wall of the skin into the arrest of the wind. Filtered air and filtered airspaces promise, then, an escape from the environment in which one must live, a small reestablishment of the body as a hard inside delivered from particulate exposure.

PERSONAL BUBBLE

In a city of semisolid airs, there is a contemporary resonance with classical Chinese medicine, where, as medical historian Shigehisa Kuriyama reminds us, the body fleetingly coalesces as an inverted pocket of air, in or out of sync with its macrocosm. "The nature of the self that slipped out of phase," like the environment of which it emerges, "was itself windlike" (1994, 33), a sleeve of energetic airs penetrating the pores and exposing them to the vagaries of a chaotic atmosphere. The hope of escaping exposure to the air, then, was the aspiration of a body safe from the environment to which it remains problematically attached and by definition vulnerable.

In a city of conditioned airspaces, architecture becomes a formal elaboration of the purified lung space. These volumes must thus be understood as the encasement of the lungs, at different embodied and architectural scales. Small spheres of atmospheric shelter are safe harbors from "the defenselessness of breathing" (Canetti 1979). Holding the lungs apart from the air, to which they are subjugated by the autonomic impulse to breathe, is an attempt at the final realization not merely of the city but also of the individual body as a conditioned and designed airspace, a mobile lung that carries its airy shelter with it like a hermit crab carries its shell. Separating air from air is an insulation of the body from its own need to breathe, a need which relentlessly exposes the body to air, breath by unavoidable breath. From the massive dome to the can of air, air design here seeks not merely, as Sloterdijk proposes, to subtly condition behavior as through an ether but also to make and hold spaces in which urban breathers might dwell and traverse. Techniques of the body include prosthesis in which a body generates its own airspace.

Masks offer a first and last line of protection in the spaces of transit that limn and separate filtered rooms. If domes and chambers rise as intermittent way stations of air, face masks allow for a last mobile pocket of protected breath in the cavity between the mask and the threshold of the respiratory tract. They are less a barrier and more a wearable technology through which a breather enters into a more salubrious cyborg relation to the air than that of a "human vacuum cleaner" (see chapter 4). A face mask's sheath of material works not simply by blocking suspended particulates but, more precisely, by making the perpetual motion of inhalation into a dynamo for a low-tech filtration system that aims to approximate the fans and filters through which rooms are made to be more inhabitable.

If, as de Certeau argues, walking in the city liberates space and makes habitation possible (2013, 162), for besieged and bemasked breathers, walking in the city is also an exposure to the slow suicide of breathing. Walking relies on a recoding of the body into a technical apparatus for its own protection. In such strange weather, a walker in the city avoids exposure and seeks shelter even when traversing through the open air of a great closed atmosphere: myself, as one airspace passing through others, splintered off into a tight-fitting bubble. But here, as elsewhere, the promise of hermetic separation falters. Suspicion lurks over whether or not a space can ever be cleaved off fully enough from another, whether or not windows can seal tightly enough against the outside, or whether the tight mesh of a mask truly captures enough. In a city occupied and remade as a series of volumes within volumes, airspaces are never well enough closed from one another; air conditioning is a response to a psychic atmosphere as much as a physical one.

Air conditioning and design remain tactics in an ensemble of environmental techniques responding to the fraught dependence of the body on air, any air. Deliverance from the air, however, is never finally a fact, and in the yawning hollows of a Beijing reengineered into breathing spaces and traced out by the amble of so many masked breathers, life and breath contend in the compromise of life in urban air.

GRADIENTS OF DUST

In a day, the view had changed. From his hotel window the day before, Dutch artist-designer Daan Roosegaarde could see the new CCTV Headquarters building, its top floors reaching impossibly toward each other to meet over an empty column of air (fig. 14). The Rem Koolhass hyperbuilding, opened just before the Beijing Olympics, was in 2008 an architectural announcement of the city's new status. While across Asia and indeed across China, cities competed to build taller and taller buildings, the CCTV Tower, if it could be called a tower, was notable for its improbable form, hanging and tilting as if to suggest a snapshot of a structure in midfall. Its form suggests a dynamism in the tension between hypermonumental heft and a floating lightness, poised as if to suggest a formal suspension of gravity. The structure, regarded affectionately in the city as a giant pair of underpants, works negative volumes into its architectural vocabulary. Teetering monumentally between eternity and collapse, it is, for Koolhaas himself, an architectural rendering of the paradoxical endurance of the Chinese state, leaning and yet never finally falling. Held in the air, soaring out of and penetrated through by a sky both open and contained, it anchors Beijing's future skyline, depicted all around it in city planning images of a future city on construction walls, each in relief against a perfectly blue sky.

And then, the building vanished. Roosegaarde, visiting Beijing in 2013, remembers that, for two days, CCTV Headquarters simply disappeared.

Figure 14. CCTV Tower, Beijing AQI 141 ("Unhealthy"), 15 September, 2017, 5:25 p.m. Photo by Micah Hilt.

Where a day before, towers pierced airspace at angles, over the next two windless days, the steady accretion of particulate matter rendered the air of the mountain-walled city basin a dense suspension. Coal emissions, car exhaust, suspended dust from an upwind desert, and the stillness of the wind, created the industrial-atmospheric condition of an architectural disappearance. Much has been and will be made of the famous pollution that has become a signature condition of China's meteorological contemporary. Simone de Beauvoir, in the revolutionary fire of midcentury communist internationalism, once described Beijing as a sprawling monochrome, a city in "gray that blended well with soils and walls" (2001, 34). From Roosegaarde's hotel window looking toward Beijing's Third Ring, this gray extended itself upward from earth and soil into smoke and sky. It is then that the optical impact of the bent tower stands in ongoing relation to the changing density of the air—that is, an architecture of negative space depends on an air that is nonetheless a variable concentration, shifting with the absence of the winds.

Where the first part of this chapter explored airspacing as a technical and quotidian practice for remaking city space in dense air, in this section, I consider two projects that address the air as a gradient of dust, each of which approaches the air through the act of condensing it into a particle solid. This is a technics of density out of which specific claims about air, bodies, and the status of breath as a human universal and a Chinese specificity are hashed out. I am particularly interested in how these projects, each addressed to Beijing's air and organized around the dramatization of Chinese air as a dispersed solid, enact atmosphere as a variable concentration that exists in multiple virtual phases. The atmosphere, that is, is not simply a gaseous medium as opposed to the solid earth or liquid ocean; it is the internal phase of a suspension whose principle technical and political feature is that it can be manipulated as a dispersed solid.

Indeed, in the two projects I consider, dust is not simply a particularly consequential substance. Its status as an aerosol interphase of solids and gases also subtends a substantial and technical limning of solid and gaseous states of matter; diamonds, bricks, and the Chinese sky appear as staggeringly interconvertible formations of the same substance at varying densities: a substantial vocabulary of dust. What they draw our attention to is a peculiar distribution of matter as it emerges as a technical principle. Aerosol interventions, then, are not simply a matter of filtrations—of removal of particulate matter for "clean" air—but an ongoing work on the concentration and dilution of the air as a massively distributed solid. That is, on gradients of dust.

I am not interested here in critiquing these projects—though there is perhaps much to critique—but rather to imagine through them as what Eben Kirksey calls "para-ethnographic objects." The principle analytic feature of such objects is that they present a way in to reorienting a problem. Also beyond the scope of this section is their strategic virality and rapid dissemination—and disappearance—in Chinese and foreign social media, itself perhaps a sudden and dissipating density of media exposures. I approach each as "a conversation piece, to facilitate unconventional ways of speaking and thinking about the issues at hand" (Kirksey 2015, 65). I hope they offer some insight into asking a question of contemporary air that approaches its political and material specificity as a suspension that toggles between and also holds together domains of architectural, respiratory, and political aesthetic concern.

INTAKE

Roosegaarde remembers, in an interview with the *New York Times*, "It's weird how the city is covered in smog and no one is doing anything" (Schuetze 2014). Of course, much was and is being done. Indeed, the conditioning of Beijing's important atmosphere by controlling sources of particulate matter—from movable factories to tightly controlled, desert dust-storm source areas—has driven a large-scale spatial reworking of governance of places now within the city's air shed. While these interventions aim to keep particulates from wafting into the air, however, Roosegaarde wonders what might be done about pollution that is already there. On the most polluted days, there is little to do but hope for the wind to flush the city's air clear, and any resident of Beijing can recount the suddenness by which the air's density can change; a brisk wind moves gray to blue. In the twenty years or so before the central government promises that Beijing's air problem might be resolved (Ahlers and Hansen 2019), Roosegaarde could gaze into the haze and wonder what could be done.

Inspired by Beijing's air, his design studio in Rotterdam announced in 2013 a collaboration with Beijing's municipal government to design and build an apparatus that could radically reduce the density of suspended particulate matter by capturing it. In its first iteration, it was to be a device that would collect atmospheric particulates by using electrodes connected to high-voltage power supplies to create a weak electrostatic field that would imbue smog particles with a weak positive charge. Grounded collecting stations could then passively capture charged pollution, effectively condensing suspended particles into a solid that could then be disposed of (fig. 15).

Figure 15. Daan Roosegaarde, with sky in particle solids.
Courtesy of Roosegaarde Studio.

Roosegaarde hoped that this technique could hollow open spaces within the air on an ad hoc basis, not by cutting out volumes or throwing out air-envelopes with physical membranes (McCormack 2018) but instead by manipulating local concentrations of aerosolized matter. Roosegaarde hoped that in the skylit space created by collecting aerosol pollution, he might "create a place where citizens, makers, NGOs, and government could experience clean air. A bubble of clean air where people could think, meet, and work together: how to make a whole city smog free."

In the suspended time of suspended particulates, the tower lives in the air by making spaces in it; rather than forcing a settling on a large scale, the towers would generate pockets of open air inside the open "dome" of the city, simulating public space precisely in the places where they already are. A technically generated caesura in the smog, in Roosegaarde's vision, would create the meteorological condition for a problem-solving collaboration between state and makers. Before the installation of the tower in late 2016, there was open speculation that this design project might become a general principle of urban air-planning, with electrostatic fields pulsing off the sides of city buildings, manipulating existing particulate loads instead of waiting for decades to breathe.

The city is in suspension, held in a time before all that the air holds has settled. In Roosegaarde's tower, the city atmosphere is a suspension in a much more literal sense, a dispersion of aerosolized solids that rise, hang, and may eventually fall out of the air. The tower, charging particles to collect them, enacts the air not as a gas so much as the medium of a massively distributed solid. His tower enacts this dispersed hyposolid atmosphere as a technical object through a technical vocabulary of intakes, collections, condensations, and concentrations. Air is not a gas but an interphasal relation between mediums and elements, one that can be induced under proper technological and political conditions into more solid states. If the air that reveals or obscures the tower reveals itself as a variation of particulate density, it is not much harder to ask how this variation can be operationalized as an aesthetic and political strategy of air management, thinning airspaces out of the great urban sky.

A successful Kickstarter campaign has funded the construction of a Smog Free Tower, a mobile architectural apparatus that uses the electrode technique to collect solids in the air. Launched in his native Rotterdam in 2015, the tower traveled the next year to the place of its conception, set up in an arts district in Beijing. In its press release, the studio announced the collection of billions of particles in its first months, while the China Forum of Environmental Journalists, an independent NGO, argues that the tower was ineffective at all but condensing air in its immediate locale at the rate of "a spoonful of salt" per hour (Huang 2016).

At the core of the tower project's conceit is its formal enactment of urban air and the dusty solids in his bag as two poles in a continuum of substantial density. An atmosphere here appears as a capacity to phase shift, one end of a gradient of dust. The slow intake of the tower means that collecting the air might be thought of less in terms of simply the passage of dirty into clean air and more as a revelation of the air as one possible phase of dust. Roosegaarde makes this clear as he holds in a Ziploc bag or in a petri dish a quantum of Beijing's air, concentrated into an ashy powder. The density of the powder is an antidensity of the sky. "You are buying a cubic kilometer of clean Beijing air," he tells the *New York Times*. For those who contribute to the Kickstarter, this air will be compressed into a dense block of soot, the center of a smog ring—here, a ring held against a vista of the CCTV Tower looming against its hazy surrounds.

Later, the language of commerce shifts the Kickstarter promotional materials into an idiom of sharing, where funding his tower is a donation not simply of money but of clean air: "By putting the captured smog particles under high pressure, we create smog free rings. And so by sharing a

Figure 16. Smog rings held with CCTV Tower, undisappeared, in the background. Courtesy of Roosegaarde Studio.

smog free ring, you donate 1000 meters cubed of clean air to the city via smog free towers." If pressurized, he suggests, the donation of clean air can become diamonds; the trick, he confides, is to press hard enough so that the compressed dust retains enough of its pigment to remain intractably a signature of the dirty air from which it comes. This language that condenses philanthropic commerce into civic air gift also introduces an uncanny and transphase equivalence between air and precious stone. The ring is a gift to the city, a kilometric airspace collected into a perfectly imperfect diamond (fig. 16).

BRICKS

Roosegaarde's tower, inspired by Beijing and finally making its circuitous way back to the city, was echoed uncannily in a performance art project released by Weibo user Jianguo Xiongdi, Nut Brother. It might be noted that "nut" is semihomophonous with *Jianguo*, or patriotic nation building; its tonal near miss resonates with other mobilization of near puns as a strategy for sidestepping and mocking state internet filters. Over the course of one hundred days, the Shenzhen-based artist released a daily image of himself in Beijing's coal-fired winter, donning a gasmask and wheeling a vacuum cleaner through Beijing's postcard sites, vacuuming the air. In interviews with Nut Brother at his studio in the southern Chinese city of Shenzhen, he told me that he blended unexpectedly into the administrative peoplescape of the city: most people assumed that he was a city worker (fig. 17).

At the end of the one hundredth day, Nut Brother took the collected contents of one hundred days of city air to the nearby city of Tangshan, where he had the dust fired into a single brick (fig. 18). Respondents in the rowdy comments section of Weibo wondered how many days of horrific air quality might it take to gather enough bricks to build a new bureau of environment in the city. As the accretion of individual days of vacuuming, the brick is a mixture of a series of snapshots in the meteorological life of the city. Composed principally of coal dust, it requires, just like Roosegaarde's electronic devices, coal to come to form. Brickmaking produces more coal dust for bricks: the vacuum runs on Chinese electricity, it burns coal to capture it. The kilns in which dust hardens into brick also belch coal into the air.

When all hundred days arrange as a tableau of mosaicked photos, the vacuum and its masked operator trace out an urban geography of the rising city, moving from historical sites to splendid contemporary architectures—including CCTV Tower—to make the psychogeography of the city indiscernible from a sojourn through its dusts (fig. 19). Beijing emerges not as the site of a political protest *per se* but as the site of an ongoing condensing. In his photo before the Forbidden City, Tiananmen Square appears as one iconic moment in a civic ambling rather than more conventional workings of the space as a direct confrontation with state power, as, for instance, in Ai Weiwei's 1997 *Study of Perspective*, a surreptitiously photographed middle finger to a Tiananmen still visible in the hazy grain of the print.

If Roosegaarde's tower—mobile, replicable, and brimming with the promise of skies compressed into gems—enacts air collection as a

Figure 17. Nut Brother vacuums Beijing air near Wangjing SOHO, day 98 of 100. Courtesy of Nut Brother (*Jianguo Xiongdi*).

Figure 18. Beijing in 100 smog vacuum-days. Courtesy of Nut Brother (*Jianguo Xiongdi*).

Figure 19. Brick: a month of Beijing's sky. Courtesy of Nut Brother.

technological opening to an atmospheric future of soaring vistas and happy collaboration in the hum and electrostatic field of machines, what does Nut Brother's brickmaking suggest? What poetics of airspace is possible in its visual contrast between the powerful vacuum gleefully taking in air and the masked face quietly hoping to keep it out? The electrical intake of the vacuum mimics what its breathing operator cannot not do: the body and the vacuum appear as doubled air-collection and -concentration apparatuses. Where Roosegaarde's diamonds imagine Beijing's air as a constellation of diamonds, distributed as a physical donation of dirty air, here the air is again compressible, pressed into hard bricks. They recall a fantasy of carbon capture and storage, where carbon dioxide might be stored as innocuous bricks, contained within a stable form and treated to slow the off-gassing that will return them to air (Günel 2016). Clearing the air is not rendering it invisible but rendering it solid.

But then there is also the matter of the peculiar impersonality of a brick, its formal nonspecificity. Argentine architect Fernando Diez writes, "The secret of the brick's form is located in its geometric regularity and manipulability. If to this one adds patience," say, the patience of a daily collection of particulate matter, the patience of a single brick each winter, "the small piece can give form to enormous masses."[9] The long repetitive toil of this brick making calls to mind the opposite of the story of the Fool-ish Old Man and his many generations of descendants, who might move a mountain by chipping away at it slowly. This story was an important part of Maoist pedagogy, in which its insistence on labor as a future-oriented practice accorded well with communist ideologies of voluntarism.[10] Here, instead, one pauses to wonder about what architecture the air bricks might make possible.

Bricks suggest a modularity and a replicability, just as the Smog Free Tower does. And yet, they make no specific claim to a future except, per-haps, to suggest that a city might be thought of as a biotechnical apparatus for collecting and compressing the air, the modularity of lungs paralleling the modularity of bricks. Nut Brother's program, if anything, is an exercise in an activity that continues without a utopian horizon, claiming no future besides the endless accumulation of the atmosphere as bricks. "What I've done," he says, "is like Sisyphus rolling his giant stone. There's no use, but it can make more people think about the issue." The atmosphere here is then a process of slow, useless conversion, an invitation to imagine the mountainous structures that might be built through the repetitive draw-ing in of the air, by the lung-like machine or the machine-like lung.

CAPTURE

Rey Chow thinks with the figure of the trap to consider the condition of being captured. Capture is a helpful opening to considering the heterono-mous and reciprocal but irreducible relations and relays that hold air in machines and bodies in air. In thinking with the trap, Chow consistently reminds that entrapment is an event of relation, an enmeshment of trap and captured thing. Prey is the effect of the trap's operation, just as the trap only comes to fruition in the violent imposition of this capture. Chow writes, "It is only in the prey's entanglement and, finally, its embodied state of captivity that the intent or the intelligence of the trap's design is fulfilled and becomes legible" (2013, 43). What is the body that can become prey, that can embody a state of captivity?

For intake devices, it may well be a particulate body that is at stake, enmeshed into the apparatus by the powerful attraction of electric charge or mechanical suction. But as an attunement to the dynamics of atmo-spheric suspension, it may ultimately implicate human bodies as them-selves part of the particulate content of wild airs. If machines capture the content of the air—that is, they enter into mutually irreducible relations of proximity and arrest—it is also that they emerge precisely in response to a physiological capture of the body by air. What remains, then, is a spiraling choreography of human and mineral particulates, dispersed unevenly in the heady suspension of city life.

PART III Continent in Dust

At my worst,
I'm thicker than water,

stranger in a profitable land.
Most days I stand at the edge
of the continent and shout

our glitterest names
in my ugliest language. We mean
what we say. I mean:

From "LINE" by Franny Choi (2020)

In the days leading up to the inauguration of a new president in the United States in January 2017, Chinese president Xi Jinping addressed the United Nations in Geneva. Arguing unequivocally for the importance of the recent hard-won Paris Agreement, which secured international agreement to keep the increase in planetary temperature at 1.5 degrees centigrade, he told of the Earth's future through a ghost story. "Man coexists with nature," he began, "which means that any harm done to nature will eventually come back to haunt man" (Phillips 2017). The statement, falling over an audience in Geneva, describes the natural world as a monstrous, spited presence, ready to exact revenge. Perhaps inadvertently, the statement in a European ballroom also conjures the *Communist Manifesto* and its famous declaration of another "specter" that "is haunting Europe—the specter of communism" (Marx and Engels 1979).

What are we to make of these two ghosts, the one conjuring the other? Straddling these two hauntings and their respective crises, world communism and planetary ecological collapse appear not simply as uncanny analogs of one another, but as spectral forces on an Earth wrought as a political and physical machine. How do these machines touch? The Chinese president positions China as the defender of the planet's geophysical system by defending the precarious political compacts that we now take for granted as the only hope for sustaining it. In the face of the ascendant US regime and its curious cabinet of professional climate gaslighters (Duca 2016), Xi's recommitment to the Paris accords was widely interpreted as a maneuver to reposition China in a central position in the global political machinery of hard-fought international agreements over climate and the planetary regulation of "atmospheric chemistry as such" (Whitington 2016, 9). China, Xi announced, would take a "driving seat in international cooperation to respond to climate change" (Friedman 2017), prospectively announcing the ascendant role of China in both the political and geophysical mechanisms of planetary ongoingness. This startling reorganization of world and planetary dynamics was noted by science commentators, who answered *China* to the question, "With the United States now planning to withdraw from the [Paris] agreement, who will replace them" (Knopf and Jiang 2017, 569)?

What lingers in this meteoric and meteorological ascent of China into this position of global and planetary leadership exceeds questions of realpolitik and strategic geopolitical maneuvering, on the one hand, and on the other, exceed the ironies in the world's recent biggest aggregate emitter vying with the United States, the world's second biggest emitter, for leading roles in the edifices of international climate politics. Outsize emissions endow the country with a political and geophysical gravity. The uncertain climate regime of the Anthropocene is reconfigured as a strategic possibility, a site to imagine the programming of Chinese global-political leadership into the continuing operation of the Earth system as such. At the same time, Xi's announcement of a China-led climate *internationale* raises complex questions about the relationship between the Chinese Communist Party and the geophysical machinery of the planet.

Incipient in this maneuver is a political reframing of the planet and its world order as variegated components in a single Chinese-led system, a political and atmospheric machinery whose gears turn around the Party-State as at once a global and planetary apparatus. If the continuing leadership of the Party has become a geophysical prerequisite for the livability of a planetary now-future, earthly survival *per se* appears as a continual effect of the leadership of the extant Chinese state as the motor of a planetary atmosphere that functions through complex political and governmental inputs over increasingly large swaths of earthly space and volume. The vengeful ghost Nature of Xi's announcement may not be a demon to be exorcised but, rather, a socialist revenant making its uncanny return: "It begins by coming back" (Derrida 1993, 11).

Reflection on planetary endurance as a socialist problematic in China has a crucial recent history. In the lead-in to the Green 2008 Olympics, Deputy Minister of the Chinese State Environmental Protection Agency Pan Yue theorized what he called "socialist ecological civilization" as a necessary and historically inevitable next stage in the evolution of world socialism, with China as its motive force. The paper was disseminated widely in state presses, and "socialist ecological civilization" has become a widely circulated slogan for all manner of "environmental" themes.[1]

In addition to setting forth the conditions of "a top-down imaginary of China's future" (Hansen and Liu 2018, 320), the notion of socialist ecological civilization that Pan elaborates puts forth a way of figuring the shift of socialism from a world-historical movement into the existential condition of planetary continuation. In the paper, Pan develops a genealogy of Chinese socialism as itself an experimental system, continually reassembling itself to deal with pressing challenges (see Ong and Zhang 2008). In doing so, he elaborates Chinese socialism as the template of a relentlessly anti-universal Chinese-led history, and in it a program for planetary survival, through the momentum of that history locked, for better or worse, in the gravity well of China's socialism.

The fall of European state socialism in the early 1990s, Pan writes, was not the end of socialism but "merely the end of one model of socialism, one experiment" (2007, 13). The end of the Cold War, he figures, offered new breathing space for the development of Chinese socialism. Its gleeful embrace of markets is here not an undermining of a socialist tradition, but an expression of its essentially anti-essentialist spirit. This embrace marks a Renaissance, a revival that continues what Pan describes as a creative and experimental tradition in the realization of a model of anticapitalist order: "China is not simply the lucky survivor of the twentieth century's crisis in socialism, but in this century, it is socialism's motive force. The future of socialist theory and systems is staked to China's success or failure" (2007, 13) and not to the inevitable unfurling of communism as a theory of world history.

For Pan, this future of socialism, threaded through the Party, turns out also to be the future of the planetary environment. Both hinge on Chinese leadership on the world-planetary stage. Because, he argues, "capitalism is the root cause of the global ecological crisis," any international consensus forged through capitalist concepts of sustainable development cannot lead to a real development, as capitalism's foundational requirement of economic expansion also, according to Pan, requires environmental destruction.[2] At a moment of crisis in liberal capitalism, Pan locates China's historical development as an alternative to both the capitalist West and the defunct USSR, and Chinese leadership as the very condition of planetary survival.

In his historiography of the future, Pan argues that "socialist ecological civilization" is to grow out of the specific features of Chinese socialism as well as specific traditions of China's traditional and revolutionary cultures. Pan's Anthropocene is relentlessly anti-universal and thoroughly Chinese, owing as much to Daoism's purported care for natural harmony as the time-tested survivability of the Chinese Communist Party. And in addition to this, it must be understood as an attempt to anticipate a future planetary machinery whose development and longevity are fully contingent on the development and longevity of the Chinese Party-State as a political and material infrastructure. The cogs of the planet, that is, turn through the motive force of China's political apparatus, figured as a component in climatic functioning.

If the Anthropocene thesis names a fretted entanglement of human and natural processes into an earth system gone haywire, socialist ecological civilization indicates the programmatic confluence of the Chinese political apparatus into the functioning of planetary endurance. In this historiography of the now-future, today's China, with its record-breaking emissions in aggregate, appears as a fleeting hangover of Western industrial civilization. The materialisms, ecological crises, and experiments with markets and neoliberalism that might be taken to characterize Reform and Opening are, in Pan's vision, an inertial and soon-to-be-surpassed configuration of politics and environment. Its principle function is to condition its own obsolescence in anticipation of the socialist ecological civilization to come. In the thesis of socialist ecological civilization, flowing seamlessly through idealized streams of deep Chinese culture and recent revolutionary fire, the Anthropocene, with its universalist pretense and its reliance on a singular species—humankind—appears as an opening for a project of nationalist ascent. The Anthropocene, with its universal subject and its tragic romance of a single unified planet, here appears as a Sinocene: a Chinese age incipient in the confluence of a Chinese civilizational future exported globally and the climatic machineries that they are reengineering, through which they must be reengineered. China and only China can sustain a changed planet through its fragile ecological disorder.

The Sinocene is, then, a figure of late socialism as planetary experiment. The Sinocene might be understood not only as one of a growing list of -cenes through which we might make sense of the history of the future but as a specific way of engineering environmental and political machineries, of drawing together the earth system and the world system through a Chinese node. Remembering that -cene means newness, it stands as just one of a parade of announcements of New Chinas that have marked the country's tumultuous modernity, from one revolution to another and, from the outside, from one Yellow Peril to the next. In uncanny contrast to notions of Gaia as a living and autopoetic planet-being (see Clarke 2020), the engineered world implicit in Chinese theorizations of socialist ecological civilization, and in global climate accords, indicates a planetary machinery that must be continually jerry-rigged through a political infrastructure turning around a great Chinese gear.

The declaration by a Chinese president of Chinese leadership on climate matters might be approached, then, as more than a shuffling of positions in an unreconstructed edifice of international power. The future of China's discrepant socialism, for officials seizing on existential environmental threat as political opportunity, appears, following Paul Edwards, to be an attempt at inducing a Sinocentric infrastructural globalism, where China's evolving political status quo will be "quasi-obligatory" (2006, 230; see Elvin 2006, xviii), a condition that cannot be rejected for fear of climate breakdown. To turn Frederic Jameson's famous citation of a mysterious "someone," it might be said, "Someone said it is easier to imagine the end of the world than the end of" (2003, 76) socialism. For in this anticipatory imagination of a geophysical machinery powered by the beating heart of the Chinese Communist Party, the end of socialism (with Chinese characteristics) would literally be the end of the world. If the future is haunted by a returned nature, it is also haunted by a returned socialism, two revenants that have become one. The specter of nature and the specter of communism haunt together, two ghosts sharing a vast machine.

6 Downwinds

RISING AND FALLING ON A CHINESE
WEATHER SYSTEM

BACKGROUND

Once a week, Dr. Matthew Swensen, a geochemist at Lawrence Berkeley National Labs, makes the drive to the top of Mount Tamalpais, just across the Golden Gate Bridge from San Francisco, to collect Chinese particulates.[1] "It's become kind of a hobby for me," he says, turning an imaginary wheel to hug a tight curve. "I have a Mini now, and it's a lot of fun to drive on the mountain roads." Mount Tam, upwind of San Francisco and without immediate "local" pollution sources between its peak and thousands of miles of open Pacific, is what his team calls a "relatively pristine" site.[2] It is one of several such sites on California's coastal ranges where his collaborators measure American air before it is contaminated by American exhausts. From the summit of Mount Tam, he recovers three cartridges of particulate samples a week—one for elemental analysis; a second for isotopic analysis, his lab's specialty; and a third as a spare for his "archive of sorts"—from a rotating DRUM impactor (RDI),[3] a scientific device that his team has placed there to collect particulate samples floating in from across the ocean.

In collaboration with an interdisciplinary team of atmospheric scientists, the goal, he explains, is to establish a provisional "fingerprint of an airmass." The impactor forcibly precipitates onto time-resolved Mylar strips, and his team's analysis will apportion this silicate coating to distinct national, geological sources. In doing so, he and his collaborators pore over dust to discern how much of American air is, stunningly, Chinese.

Swensen and his atmospheric scientist collaborators approach airmasses as inbound landmasses. Their meteorological geology is a folding of wildly divergent timescales across the hemispheric geology of dust events. It is streaked with the imprint of China's Reform and Opening in the effluents that dust clouds entrain into their globetrotting itineraries: nitrous oxides, and ozone precursors; coal soots, the pyro-geological signature of China's rise. In our conversation, Swensen estimates that at times a full 40 percent of particulates collected in this first American air are demonstrably Chinese in origin. This portion spikes in the early spring, when dust season in China scatters Chinese geology on seasonal airstreams that ply the hemispheric atmosphere, to settle, among other places, in the rotating hold of the lab's DRUM impactor. In 2001, a historic storm transported mineral dusts from Asian deserts over the continental United States in an amount "comparable to all U.S.-based sources" (Jaffe, Snow, and Cooper 2003, 503) of particulate matter.

In their various attentions to these aerosol airmasses, they pose the vast distances of the Pacific as a riotous tangle of windy highways, channels for continental drift in long-traveling Chinese earth. Computer models have suggested for decades that Asian aerosols can move across the ocean. Supplemented by space-borne and ground-based observations, dust aerosol models have linked the deserts of Inner Asia to North America in a seasonal weather system in terrestrial plumes (Guo et al. 2017). Swensen, with a mischievous smile, prefers not to rely on models alone. "I prefer to get my hands dirty," he says, quite literally by playing with the particle traces impacted onto the RDI's strips as a desert pulled out of the sky. The rise of China's economy registers as a wave of particulate matter, a cloud of dispersed solids and entrained gases that bears the distinct geochemical signatures of decades of the Chinese style of development. Impacted dusts route America's Pacific Coast into the trajectory of a seasonal inbound continental airmass. The edge of the continent has become a line in the wind, where the meteorological fallouts of the Chinese economy make landfall.

"We've been waving our hands for a long time about long-distance transport," states Swensen, "and what's great is our lab was actually able to see and quantify transported material." Seeing and quantifying this material means characterizing the American spring as a light rain of land at the Pacific Coast and parsing out the contents of the air into a distinct geochemical type: a meteorological stratigraphy takes shape in the analysis of aerosols. The team cumulates dust samples from the Mount Tamalpais site and other "relatively pristine" locations into a "China-in-the-US data set." After isolating out more local aerosols that may end in the impactor's sample, like sea spray and salt, the team compares the China-in-the-US figures with particulate samples collected during the same time periods from sites just downwind of Bay Area conurbations. Subtracting one from the other yields a "US-in-the-US data set."

America, in the interplay of data sets, is an excess over the trace of China that comes to stand as a baseline of incoming air. It is also what is left over after that Chinese baseline has been ruled out.[4] The push to develop methods for apportioning California's air into distinct local and foreign parts escalated beginning in 2015, in response to tightening federal regulations on surface ozone concentration under the Obama administration. Air quality management districts in western states appealed to the US EPA, arguing that foreign ozone and ozone precursors, per Article 179B of the Clean Air Act, were often enough to push western air quality over regulatory thresholds.[5] Article 179B exempts air management districts from responsibility over so-called background emissions, those emissions that can be demonstrably sourced outside of the territorial airspace of a given jurisdiction, or consigned to legally "natural" events like wildfires or downwelling of stratospheric ozone.

For air quality regulators with tight ozone budgets the sciences of aerosol transport and techniques of particulate apportionment present a chance at moving the bar on the quantum of pollutant that domestic air districts are responsible to address. Suspicions over China's environmental footprint play out in impactors, relatively pristine sites, and research labs across California, in an insistent attention to traces of springtime dust, phased into silicate coatings on Mylar. Regulation depending on methodological innovation remaps the aerosol makeup of dust samples into a cartography of fixed national sources. In this gambit, with its regulatory contortions and its bricolage of sciences and strategies, it is China, and not *nature* as such, that is the baseline against which America will take shape as a super-added atmospheric contribution.[6]

Through trace dust samples, they enact dust events as a seasonal connective tissue that, against the backdrop of transpacific geopolitical ties and tensions, revises China as a geological constituent of domestic air in its most physical sense. As impacted dusts unfurl, through their analysis, into geochemical signals of Chinese land, of burned Chinese coal, of the emissions of Chinese factories, they are learning to understand China as a weather condition, and the United States as a point in the far-flung meteorology of a place upwind. Lead isotopes, in particular, have been proposed by interdisciplinary meteorologists as a "tracer for airborne particles within the growing Asian industrial plume" (Ewing et al. 2010, 8911).

Even as local groups in California deride this sudden importance of China upwind as an opportunistic ploy at outsourcing emissions and thus local responsibility to a convenient upwind villain, scientists and regulators conjure dustflow as a phase-shifting medium of Sino-American relation. International relation is reshaped along the phases of earth and air, from deserts, to dusts, to gray bars, and finally, in a labyrinth of regulations and good enough scientific work, into legal baselines. The transpacific airstreams that NASA has called the Pacific Dust Express are a distinct skeleton of planetary relation through waystations in a geography of traveling Chinese effluents.[7]

At UC Davis, where the aerosol "archive of sorts" is housed, Dr. Chen, who designs, builds, maintains, and customizes the rotating DRUM impactors, moves an old microwave and several cases of scientific instruments. He is a lab technician responsible for maintaining and customizing the nine RDIs currently in use, including the one at the Mount Tamalpais outpost. This lab's "archive of sorts" is buried at the end of this labspace Tetris. He blazes the two of us a body-width trail through the haphazard office to a card catalog cabinet that holds neatly labeled Mylar strips retrieved from RDI's arranged at various sites up and down the coast. The impactors drag air through their warren of pipes and drums, inducing a dust storm in reverse to compile an aerosol record of incoming airmasses as a topography of minute landmasses.

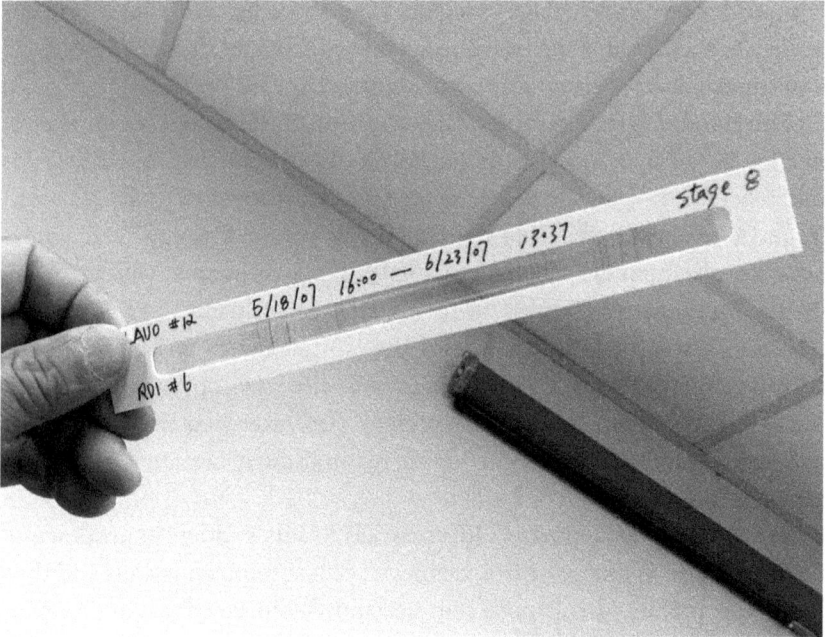

Figure 20. A month of China in the air. Rotating DRUM Impactor sample strip. Photo by author.

Holding a strip against the light, he sees a month of China in the sky (fig. 20). Each faint stripe of gray is a particulate sample collected over two days. Dusts forcibly settled onto Mylar are ex-land and then ex-air. The China reconstituted in the machinery's hold presages aerosol transport as a medium of hemispheric, trans-oceanic entanglement. They manifest Sino-American relation in the idiosyncrasies of a modern weather system. Scanning for drifting particulates, scientists and regulators stake their roles in a meteorological drama that unfurls along an idiosyncratic geography, charted in dust and the worlds that form through it.

Through dust, they conjure the vast distances of the Pacific as a fluctuating tangle of atmospheric highways, billowing open and contracting with the seasons. They enact China's rise not in the increasingly bellicose languages of trade wars and great powers, but also as the foundational strata of American meteorology. China's presence matters, as it registers in labs and legal provisions as "changes in the ambient composition, density, and texture" (Gordillo 2018, 58), and in this context, US emissions stand as an excess over the Chinese geology that is always already there in fluctuating abundance. China and America fade into view through geochemical analysis the mapping of aerosol transport and transition across a delimited strand of voluminous planetary space. By analyzing it, decomposing it, parsing it into countries and origins, forcing it into clauses and exemptions, they generate visions of transregional connection that can be neither deduced from the wind nor the law alone. And in this, they are one act of a meteorological drama staged across oceans and oscillating through phases of dust.

———

This chapter learns from the DRUM impactor in the search for a contingent meteorological architecture through which a more-than-human world-system might come to form. In the curves of its machinery and its slowly rotating chambers, the impactor displaces long-distance relation into a matter of precipitations. This mechanical deposition of particulate matter out of air onto Mylar effectuates a dust storm in reverse. It configures continental geochemistry and meteorological territorialization as a matter of falling particulate matter. And for the impactor and the experimental system constellated through its steady churn of incoming aerosols, the air is not an emptiness. It is a freight of particulate traces that come from *somewhere*, and which become experimentally and politically salient as they are precipitated, forcibly phase-shifted through impaction into a substance through which hemispheres, centuries, and the geopolitics of Sino-American entanglement are reconfigured.

How do weather systems and world-systems pattern into one another? The DRUM impactor and its scientists pore over inbound dusts to pose China, awash in the fumes of its coal-powered, dust-shocked rise, as a constituent of a meteorological relation. In the impactor's spinning chambers, the world-system that has been offset by East Asian economic miracles subtly transmutes into a weather system. The flows of commodities and the financial networks anchored by cities along the Pacific Rim find their geography mimicked by hemispheric airstreams. I attend to how "Chinese dusts" and their uptake as "foreign background" downwind might give us insight into how aerosols and airstreams might rut out new time-spaces of relation. These dusts and the winds they reveal as vectors of relation fold political and meteorological geometries into one another, as when international relation suddenly bends into the shape of the wind. Together, they offer a site of entry to a modern weather system, one that gathers and stimulates divergent political, material, and environmental becomings in an atmospheric scale that attend to far-flung connections.

We return to the dust event that opens this book's introduction. Here, I ask that we attend to dustflow not as a straightforward matter of material relation, but as a material-semiotic problem through which the problem of China, for its downwind neighbors, is rewrought as a question of strange weather. In 2001, several months before China joined the World Trade Organization, two massive dust storms formed over Inner Mongolia, on the Gobi deserts of China's continental interior near the Mongolian frontier. They passed over Beijing and left Chinese airspace to reach the Koreas a day later. They next blew over metropolitan Seoul in thick pulses of dust, heavy with the coal soot and industrial emissions that the storm entrained into composition.

By what aerosol scientist Patrick Chuang described to me as a stroke of good luck, an international team of atmospheric scientists were already gathered to catch, analyze, and model this historic aerosol emission. Scientists gathered by the International Global Atmospheric Chemistry Program were already stationed just downwind of China, directly in the path of the storm. They conducted research from ground stations on Korean and Japanese islands, from airbases that sent three research airplanes on dozens of flights, and monitoring ships as part of ACE-Asia, the Asian Pacific Regional Aerosol Characterization Experiment. They establish that Northeast Asian aerosols were changed by air pollution in many ways (Huebert et al. 2003), making mineral dusts a vexing conundrum for atmospheric chemistry. As storms leave Chinese airspace, they are recharacterized as Asian aerosols. This invocation of Asia is a question of the multiple possible and interacting ways of characterizing Asia and imagining inter-Asian relation.[8]

Twelve days after the first cyclone in Inner Mongolia, upwind of Beijing, the *Denver Post* reported a column of haze rising eight miles thick over the American Rockies. It dubbed these dusts the "latest import from China"

(Schrader 2001). In this alarm over Chinese imports, the newspaper renders the open Pacific as a theater of operations for a relentless incursion of economic and meteorological invasions. Dust, here counted as one among a barrage of other Chinese imports, indexes an irresistible relation to China as the origin of an exogenous, traveling threat. Dust appears as redundant, a mocking planetary iteration of a reorientation of global relations around a Sino-American axis, measured in outsourced production and registered affectively in the exhaustion, anxiety, and "excitements" of supply chain capitalism (Tsing 2009, 149). Scientists, for instance, have argued that the outsourcing of US manufacturing to China has resulted in worsening air quality in western states as Chinese pollutants and manufacturers alike make their way back to the United States (Lin et al. 2014). Such tracing specifies chains of relation that make dust and commodity flows into surprising commodity *anti*-fetishes, especially as American economic anxieties fixate on all matter of things Made in China. Trans-Pacific relations are continually and anxiously mapped and then revised into vectors of threat pitched toward America, a storm-battered fortress, its moat an ocean wide.[9]

But not all imports are the same. Winds are not commodity chains, and the planetary itineraries of an ocean-crossing storm are not the integrations of World Trade and its exasperating drama of trade wars. This phantasmatic isomorphism of meteorological and political economic threat stages a fantasy of subsuming one into the other, and each into a Sinophobia drummed up in unstable times (Billé 2016). Collapsing them into one another is to read in emergent conditions and phenomena only more of the same. As the people and machineries that populate this chapter and its globetrotting dust-streams implore their colleagues to recognize, orienting to dustflow opens the reheated languages of international relations into a series of experiments in downwindness. Wind "enters here as an interruptive force, awakening air and rousing it from stillness" (Howe 2019, 2). So roused, the world-systems that shift around China's rise are unsettled into weather systems that pass over its unstable, weather-prone continent.

While previous chapters have lingered in the arrest and suspension of deserts and dust clouds over China, this one traces out the continents in dust that breach the vertical containments of territorial airspace to travel over the northern hemisphere.[10] Its field sites and attentions move with volunteer teams from China and South Korea who cooperate to raise a Sino-Korean Friendship Forest. They seek to offer, in rows of trees that protect both Beijing and Seoul from storms, a vision of Asian cooperation and shared fate shaped in the entangling of atmospheric and geopolitical connections. We then continue back and forth across the Pacific with the dust traces that occupy American atmospheric scientists and regulators. Tracking the trajectory of a Chinese dust storm, for these actors, reveals airstreams as important seasonal geo-atmospheric architectures of relation that both echo and displace more readily available ways of thinking and legislating relations between countries in the shadow of China's rise. The chapter closes by returning to the far shore of the Pacific, where dusty stews of isotopes hold open the possibility of another planet taking shape.

Downwindness, in these cases, is not a ready-made material ethics or political principle that can be discerned from the hard fact of fluid relation. It is instead a medium of speculative political meteorology that offers concrete scenes of practices in which sharers of dusty airstreams enact the political as a planetary emergence. How is downwindness generated as a set of technical, legal, and meteorological concerns? How, in practice, does the geographical fact of downwindness pattern into matters of relation, ethics, and entanglement on a planet unevenly drawn into choreographies of dust? Those in the course of Chinese dusts experiment with downwindness as a matter of geographical accident that is becoming more and more significant in a time when geopolitical and economic entanglements with China are emphasized, ironically, in the texture, contents, and dynamics of the planetary atmosphere along airways like the Pacific Dust Express.

Dust, in this sense, stages questions of the political and meteorological in the curious manifestation of China as a profound source of anxiety in both geopolitical and geophysical terms. Moving between northern China, the Korean peninsula, and the West Coast of the United States, I explore how China is enacted multiply as a meteorological, political, and relational matter. Dust offers a medium through which I respond to Jerome Whitington's call for a "situated, non-totalizing view of the planetary" (Whitington 2020). Aerosol transport, for scientists and regulators, opens a view on a planetary imbroglio of continents and jet streams that

is planetary but decidedly not *everywhere*. This geography of airstreams
and geochemical freights varies with windspeed and the size of particles.

I track practices that demand a reimagination of what it could mean to
be downwind of a Chinese continent. These do not forego the questions
raised by the increasing globalization of the Chinese economy, but seek
to articulate a possible language of relation in the commingling of planetary and global predicaments. The various commodity and particulate
imports that materialize China downwind provoke languages of relation
and entanglement that neither erase difference nor fall back to terrestrial
questions of national sovereignty that are in any case outmoded by the
weather itself.

How might a hemispheric airstream suspend and resettle a world-
system that is itself in flux? In Immanuel Wallerstein's notion of a world
capitalist system, capital generates historically contingent architectures
of relation, opening scales of analysis that are inherently relational and
structured through its dynamics. Wallerstein's notion of a world-system
insists on foregoing the territorial nation-state as a unit of analysis and
comparison, emphasizing "the historically specific totality which is the
world capitalist economy" (1974, 391). It can be traced in its own dynamics
of "rise and future demise." Capital, in this account of the world-system,
coordinates divergent and disparate countries and political formations,
materializing in an internally variegated totality that can be traced from
its historical rise to its inevitable fall. Its world-system takes shape in
the way capital generates positions, creating an architecture of emergent
differences in a "grid of exchange relationships" (1974, 397). Despite its
totalizing demand, it does not imply that connection is tantamount to
the equalization of its constituent parts. For anthropologists, the capitalist world-system and its economy are a way of articulating the materials
and processes through which things are gathered and diverge. Through it,
the globe sprawls into a world of cores, semi-peripheries, and peripheries,
each a moment in the life of capital.

Relations of up- and downwindness require that location is sequential and relative. In the circulation of the atmosphere and its revelation
of flat space into geo-aeolian vectors every downwind is likewise upwind
of elsewhere. For a dust storm is not simply the blowing open of planar
politics into volumes and a politics of verticality, an observation that holds

in relation to many kinds of environmental process (see Billé, ed. 2020). It also demands a novel political spatialization that attends to its processual dynamics, to a time-spacing defined by lags, anticipation, and the anxious scanning of the sky as a medium of meteorological and political proximity.[11] To explore up- and downwind relations through these aerosol forms demands that the relentlessly specific and yet elusively more-than-territorial trajectories of particulate flow must be approached as a formal principle of meteorological relation.

In suspension and processes of aerosol formation and settling, world- and weather systems contend and oscillate. Sometimes one momentarily eclipses the other (Strathern 1988). Dust, as a material that flows across well-established national boundaries, orients those downwind of China toward specific planetary spatializations that entrain longer histories of multilateral relation, suspicion, and empire. Dust is a planetary substance that refuses the ubiquitous all-space of the planet *per se*. It allows us also to see these points in the wind as part of a dynamic totality out of whose properties—volumetric, timeful, interphasing—political geometries are patterned into permutations whose patterns cannot be discerned easily from existing political maps.

Cymene Howe writes that "wind in itself cannot be held," even if its force can sometimes be harnessed. "It is elementally loose" (2019, 12). If a world-system is a way of accreting and differentiating relations with capital as a medium of entanglement that generates its own scales and entanglements out of historically contingent conditions, this looseness of border-crossing winds also makes it a curious architecture of relation, that spills over the awkward, landlocked accountings of relations between fixed territorial states. Their smooth looseness is at once an affront to the bumpy formalisms of international relations, but also a provocation to other morphological principles. The wind introduces the trajectories, vectors, and sequences of aeolian position in frictive tension with the circulating exchanges of capital and what Aihwa Ong has called a "container concept" of sovereignty that locates political formations in spatially discontinuous, internally homogeneous blocks of land (2006, 98).

The remainder of the chapter lingers at points of aerosol transition dispersed across three countries. We follow dusts in their permutations and slipping across phases. Across points in the wind and phases of dust, problems of being up- and downwind subtly reshape international relations into questions of relative location, vulnerability, and sequence. The Chinese desert and the Mylar strip on a Californian coastal range describe a before and after of a dust cloud, thematized around the geo-meteorological entailments of the wind as an earth-moving, world-making force.

ASIA IN VECTORS

When former Ambassador Kwon Byung-Hyun speaks of the wall of trees in Chinese Inner Mongolia that has occupied him since retirement, he recounts a single storm. He recalls that in 1998, while he was serving as the first South Korean ambassador to the People's Republic of China after the renormalization of Sino-Korean diplomatic relations, Beijing was caught in an unusually severe dust storm, the first of many during his tenure. "The gales that brought in the sand and dust were very powerful," he writes, "and it was no small shock to see Beijing's skies preternaturally darkened" (Kwon 2012). The pall of dust in the city affected his breathing, as it did for many Beijingers, who were advised by state agencies to wait indoors until the storm had passed.

The next day, at the South Korean embassy compound in Beijing, he received a phone call from his daughter, still in Seoul. She complained of a bout of Yellow Dust, Korea's infamous *hwangsa*, over the South Korean capital. This time, the event was strong enough that it led to a spike in hospitalizations and the mandatory closure of schools across the Korean peninsula.

"I realized she was talking about the same storm I had just witnessed," he remembered in an interview. "I saw for the first time [in 1998] that we all confronted a common problem that transcends national boundaries." The dust was not a Chinese problem, or not only. He understood, in the offset parallel lives of himself and his daughter in a transborder dust event, that "the yellow dust I saw in Beijing was my problem, and my family's problem" (2012). It was at that moment of epiphany, precipitated in aerosol relation, that he came to a dawning realization about China and Korea, the two countries whose friendship he had spent his adult life building. They were not simply regional neighbors split by long-standing political divides and bound by the economic interdependencies sparked by China's Opening. China and South Korea share no land borders. But in a storm, they were, for better or worse, sharers of an airstream, a first and second in the itinerary of a storm. Kwon realized that he and his daughter, divided by a sea, were nonetheless consecutive points in a spring wind.

Kwon later reflected on that moment, stretched over days, as a turning point in his life. In a region characterized by historical enmities, and whose twenty-first-century international relations are delicate, to say the least, Chinese dust traced the contours of a meteorological Asia superimposed in a turbulent geography of airspaces and dust trajectories over China, the Korean peninsula, and, after another day, Japan. As Kwon worked to orchestrate reopening political ties, the dust storm made atmospheric channels relevant as another modality of Sino-Korean relation. Geopolitical and geo-meteorological densities seemed to mirror one another. Beijing and Seoul were not simply two capital cities of states at the threshold of their political reopening. They were also two points in a course made viscerally and geologically evident in the movement of a cloud of particulate matter.

While still in office, Kwon understood that land degradation in China would inevitably become an Asian problem, reaching as far as the idiosyncratic surge of dust storms. While land degradation in China's interior would affect, was already affecting, China's downwind neighbors, he continued to receive lukewarm responses to his call for joint action on dust storm control. "The Chinese I spoke with explained to me that deserts are a regional problem," he remembers. "From the Korean perspective deserts were China's problem" (2012). Meteorological integration through dust was proceeding apace of political and economic entanglement of northeast Asia's national economies.

Kwon's demand for bilateral cooperation is also a practice of fitting states into the wind, of displacing Asia through *wind-sand*. In land degradation and its concomitant continental drift by Asian winds, nation-state divisions shifted, in his reckoning, into an uncomfortable meteorological kinship. This kinship, between parents and children and between regional neighbors, would require a reformatting of Asia as a meteorological continuity, and reimagining international relations as part of the workings through which *wind-sand*, from desert to dust storm, must be addressed.

The transborder geographies of dust flow, Ambassador Kwon insists, are not a matter of upwind sources and downwind sufferers, but a matter of confronting a common problem. For one thing, as his son and other Korean forestry professionals involved in anti–dust storm work emphasize, dust storms can be understood as a meteorological signature of a changing political economy, especially as dirty Korean production and industry offshore to China. This sentiment was echoed in interviews with other Seoul-based tree-planting organizations with Inner Mongolian planting sites who invariably insisted that China and South Korea were, respectively, a "dust source country" and a "dust-exposed country."[12] This fact, however, was for these organizations completely inseparable not only from regional economic interdependence, but also from what was often framed as an exploitative outsourcing of dirty production to China.[13] The complexity of dust as a foreign problem thus at certain points was framed as a question of the reimportation of domestic particulate matter from abroad.[14]

The meteorological and economic integration of the region entertain a complex relation with one another. Tracing dust flows, their Asia cannot be singular or self-evident. As ambassador, Kwon's exhortation toward dust and politics as coinciding questions does not dispense with diplomatic concerns that occupied his professional life as an envoy working to strengthen regional bonds. Rather, it expands their scope, allowing Asia to multiply into a space constituted in an entanglement of manifold regionalizing forces and substrates, whose interaction does not require or allow final resolution. "[A]n essential, ontologically pure Asia as self-evident, self-sufficient, and self-made" (Yan and Vukokich 2007, 212) would splay into new configurations of space, time, and relation.

It was at that time, on the eve of his retirement, that he decided he would dedicate the remainder of his life to afforesting Chinese deserts, setting the ambitious goal of raising "a billion trees in the desert" in a Sino-Korean friendship forest (*Zhong-Han youyi lin*). The friendship forest would physically embody relations between two countries as a matter of densifying political ties and the thickness of border-crossing airstreams. Its planting and growth would express, in the changing landscape and airscape of a northeast Asian dust shed, a biographical shift that had already happened for the ambassador: that China and Korean relations must be built in multiple domains, converging in an overarching hope of regional amity.

As first ambassador in the 1990s, Kwon began to elaborate a complex regional vision of Asia as a meteorological and geopolitical space. For Kwon, normalization and a crisis in dust braided into an emergent vision of Asia as a convergence of multiple forces of regionalization. Since 2001 Kwon's organization, "Future Forest," *Mirae Sup*, has led groups of Chinese and Korean student planters to a planting site on the eastern edge of the Qubqi Desert in Inner Mongolia. This site has strategic import for the organization: the ambassador's political connections from his decade of diplomatic work allowed the group access to a site, materials, and labor through the cosponsorship of regional China Communist Youth Leagues in Inner Mongolia; with an overnight train from Beijing, the site is accessible for groups arriving from Beijing and Seoul.

But, as the ambassador's son, who now heads the organization, intimates in our conversation in his Beijing office, and again on a spring planting trip, the planting site is most important as a point in the path of a storm. Gesturing at a map of Northeast Asia in a PowerPoint Presentation he has prepared for Korean NGO groups, he points at a Seoul where a series of arrows designating dust-transporting airstreams converge. Tracing these arrows back, through Beijing, he hovers at the point they have chosen as a planting site in the Qubqi Desert. "These arrows are the lines that carry *hwangsa* through Beijing to Seoul." At the planting site, there is a vertical line that indicates the positioning of the Friendship Forest they aim to build, erected perpendicular to the channels of the wind. With a downward thrust of his hand, like a razor severing the dust's thread, he says, "This wall will be a first line of defense against the dust, for both Beijing and Korea."

The rendering of dust in flowing arrows (fig. 21) enacts Asia as the backdrop of a repetitive meteorological process, while centering storms themselves as a regionalizing principle. It is these arrows in their multiple iterations that are crucial to grasping the irreducibility of the spatial reckoning through which Korean antidesertification groups present the relationship between various places in China and Korea to the conventional

Figure 21. Future Forest's *sand-wind* Asia. Captions, clockwise from upper left: "Strong Siberian wind"; Wind Hole "Helan Range: a large gap of about 200–300 km"; Qubqi: "Conveniently accessible site with abundant groundwater"; "Most important source areas for Beijing region dust storms." Korean under "Sand Storm" reads *hwangsa* (yellow dust). Courtesy of Future Forest.

international relations imaginary of earthbound territorial sovereignties rubbing up in horizontal space.

While Future Forest's wind and dust maps conjure a China and Korea that appear as the background to the movement of the storm, the maps through which other organizations in Korea explain their involvement in China illustrate the entanglement of political and meteorological relations more directly. The Seoul-based shrub-planting group Ecopeace Asia defines the need for Koreans to protect themselves from Yellow Dust by addressing the problem at its root in China, because "we [in Korea] cannot cover the sky with dust-proof blankets" (Park 2011, 3). The organization uses another map of dust routes to present a Korea embattled by winds from China. It promotes a sense of an atmospheric Korea that goes far beyond its borders, systematically subsuming China's landmass into a fount of potential Korean weather. In this vision of Asia, China's mobile land and multiple conventionalized routes of dust are an unfolding of

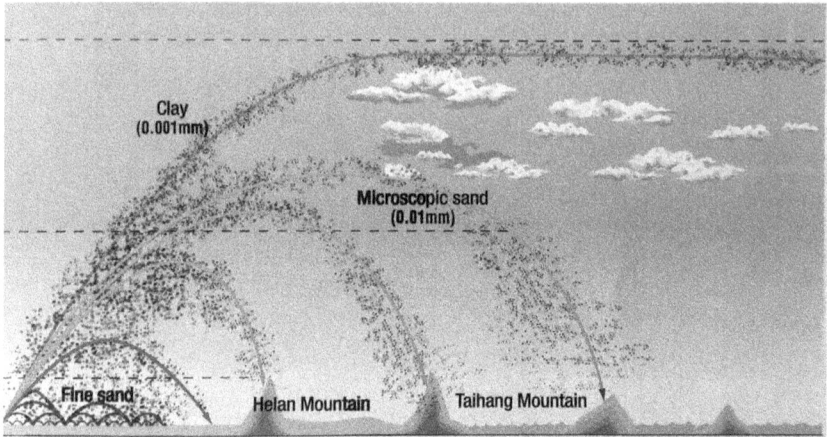

Figure 22. Asia by dust range, graded by particle size and constant wind. Courtesy of EcoPeace Asia.

Seoul's airspace upwind and backwards in the time of stormflow. Space has been enacted as distance, which in turn has become a function of hypothetical particles, resolved across a gradient of sizes, and all tracked to the steady speed of a hypothetical wind.

Polygons and arrows contend as different modalities of spatialization. The Asia that emerges out of a sense of Korean meteorological vulnerability to its upwind neighbor is tracked to the arrows that skim their way across it: it is an Asia composed not out of landmasses but vectors, lines with magnitude—velocity—and, crucially, direction (fig. 22). Territories are gathered as points in pathways of dusts, projected through particle sizes. Places become a scattering of what Alix Johnson calls "on-the-way places" in which countries become an aerosol "node, relay, or in-between" (2019, 84)

A vectoral conception of Asia reorganizes spatial relations wholesale. Places can thus be thought of sequentially in space and time, where Beijing and Seoul are not simply two places in abstract space, but a before and an after, a relative first and a second in an order. It is this vectoral quality that generates and differentiates positions internal to the dust storm's path, and where specific places are sorted out and then related to each other relative to wind's motion. In this way, places are not related as points on a plane, as in the diplomatic relations between sovereign states that ex-ambassador Kwon served to build. Rather, the dust storm path offers its own linear-spatial template wherein spaces are temporalized, measured in days of speculative storm-time.

no one knows
how to be
loving and also
hope the wind
in a certain
and not another
direction will blow
"Poem for Japan" by Matthew Zapruder (2012)

Future Forest's stormwall is a cut into the wind and a living symbol of a friendship along an atmospheric corridor. It is, in the glitzy ceremonies and planting trips that they take each spring, a friendship forest, marking both political normalization and also relentlessly emphasizing Chinese dusts as a medium of shared meteorological fate, rather than an occasion for atmospheric blame. The location of the forest project places it upwind of both capitals, allowing the wind to sort these cities into a condition of shared and distributed vulnerability. By tracing the motion of the wind, he argues that dust is not a problem of what Chinese irresponsibility and environmental challenges have inflicted upon its neighbors. Rather, it challenges those arranged along the vectors of an Asia in dust to invent, out of particulate matter and geopolitical instability, a new ethics of relation.

There is widespread public derision in South Korea of dust storms as proof of a Chinese failure to contain its bad weather at terrestrial borders. Newspapers regularly report on yellow dust, *hwangsa*, over South Korea as Chinese in origin, "carrying with it fine dust particles that contain various pollutants, including carcinogens" (*Korea Herald* 2015). In 2015, Greenpeace sought to charge South Korean officials to take responsibility for domestic air pollution, arguing that "from 50 to 70 percent of particle-laden smog, which is also known as PM2.5, is generated within the country," and disclosing that 30 to 50 percent% of particulate pollution originated in China (Jung 2015). When storms reach Korea, they contain not only dust but also the coal, chemical, and exhaust signatures of China's economic miracle. What a storm reveals in those windy circuits is that China is already Korea's atmosphere. It passes over the country and settles over its land and into its bodies.

Against this, the ambassador offers the friendship forest as a rebuke for those who collapse international airstreams into the reheated languages of international relations, looking to the air only to rehash the enmities and suspicions of the Cold War across northeast Asia. Through their work planting trees, Kwon and his cohorts offer "friendship" as a way of orienting political and meteorological relations in this air that skewers two countries, that charts an Asia, meteorological and geopolitical, as a confluence of vulnerabilities and entanglements, dangers and possibilities, with a history as concrete and complex as the wind that holds territory together like tendons.

That the ambassador's tree-planting project could happen at all relies on a delicate work of drawing on connections and mobilizing them toward meteorological friendship. The forestry program was bolstered by both the broad political support of greening projects in China, with its own standards for planting techniques and choice of planting stock, and the cooperation of local villagers and cadres who were responsible for the vast majority of actual planting and maintenance of the forests. and decades of Korean forestry expertise in the aftermath of the South Korean "forestry miracle" of the 1960s and 1970s, "born of the particular circumstances of the ROK at the time—its Cold War context, its export-oriented growth, its authoritarian politics," and the legacies of Japanese colonial land management (Fedman 2020, 232; see Fedman 2018). This expertise in reforesting a devastated South Korea in the aftermath of the cease-fire agreement that paused the still-ongoing Korean War had, by the early 2000s, become part of a broader realignment of South Korean forestry expertise with international projects. This included, in the past decades, South Korea's forestry administration becoming a leader in antidesertification afforestation expertise. The country has hosted multiple Conferences of Parties, meetings regarding the implementation of the United Nations Conference to Combat Desertification, despite South Korea having no deserts within its borders.

On the ground, the annual seasonal work of friendship forestry requires the continuing political connections and the significant goodwill that the ambassador and his cohorts had accrued in decades of diplomatic work, passed on to and nurtured by his son. It required cosponsorship from political organizations in Inner Mongolia, renewed in a ritual calendar of banquets, glowing press conferences, and smiling photos with red banners with bilingual couplets that celebrate ongoing work. As the organization runs in a gray zone as a nongovernmental organization that was nonetheless founded by an ex-ambassador and projects soft power through cultural exchange, it also requires political and financial support from the South Korean government and *chaebol*, Korean state-sponsored corporate conglomerates. In the organization's constant work to renew the financial, personal, and quasi-diplomatic relations that allow the forest to grow and its airstream to build international friendship, the organization must continually emphasize that dust from China does not raise Korean grievances, but rather, stands as a planetary injunction that two countries recognize their shared purpose.

Each spring, the ambassador's organization convenes a Green Corps, a group of Chinese and Korean university students that arrive at Future Forest's planting site in Inner Mongolia for a week of plantings, cultural exchanges, and ceremonies. Multiple Seoul-based groups, including Ecopeace Asia and Future Forest frame their missions not only in terms of the technical and logistical work of planting windbreaks, but indeed building a human resources infrastructure of future environmental leaders gathered by common cause through aerosol transport from China to greater East Asia. In Future Forest's planting trips, they anoint these binational teams of planters as a select group of students handpicked to become leaders for a future Asia. They will build relations through the shared work of planting in the desert, which is part of an itinerary of other group-building events like banquets with talent shows, and group trips to a *jjim-jilbang* Korean-style bathhouse that has opened near the Qubqi Desert to accommodate Korean groups coming and going from the planting sites.

They meet first in Beijing, then take an overnight train together to Baotou, a coal boomtown in Inner Mongolia, where I meet them before they head toward the Qubqi Desert site in a flotilla of tour buses, one of which strands itself in a road barely marked from the loose sand all around it, its tires spinning ineffectually and throwing up clouds of dust. The ambassador and his organization inaugurate these binational planting teams as the stewards of a fragile Asian friendship. In these trips, the students trudge across dunes, build windbreaking straw grids, and plant trees; they continually describe the shared exposure of Chinese and Korean airspaces to continental dust as a crucial opportunity for building relationships and infrastructures for international cooperation. In the chartered bus where I ride with Korean college students who major in economics, environmental studies, Chinese, and English, who laugh at good-humored translations between three languages that replicate both the geopolitical and meteorological connections of an Asian Pacific in conversation.

Friendship is a term oriented in shared and distributed exposure. It is an opening in the search for a political and ethical vocabulary for being-downwind, and for inhabiting a vector of *wind-sand*, with its internally heterogeneous totality. The forest allows for this friendship, first, in that its very position—the outcome of negotiations in a field of political, meteorological, and logistical considerations—is upwind of both Beijing and Seoul. This locational grace allows for downwindness, a matter of relative position, to be something that the capitals share. They fold the ethical injunction of this choice, its small geographical kindness, into the agency of the dust itself. Its materiality can be mined for atmospheric configurations other than blame.

Friendship along an airstream is a demand to see the rush of dust-arrows to Seoul (fig. 23) as a question of sharing unequally in a modern weather system. It does not demand an immediate accounting of meteorological damages, a stormy version of what Julie Chu has described as a "politics of destination" in the transnational itineraries of other Chinese things (2010). Rather, it is an ethical opening without resolution, one that asks us to grapple with the matter of atmospheric relation, both as a facet of thickening geopolitical dependencies and as something wildly irreducible to their zero-sum frame. For the ambassador, whose professional life as a diplomat has been dedicated to achieving it, friendship is a political reset, a demand that the two countries strive toward relationships that catch up with the wind, fit for the meteorological reality at hand and in a future marked by both regional interconnection and regional ecological instability. Friendship in this sense names a relationship that does not yet exist, a telos unreachable as long as only diplomatic

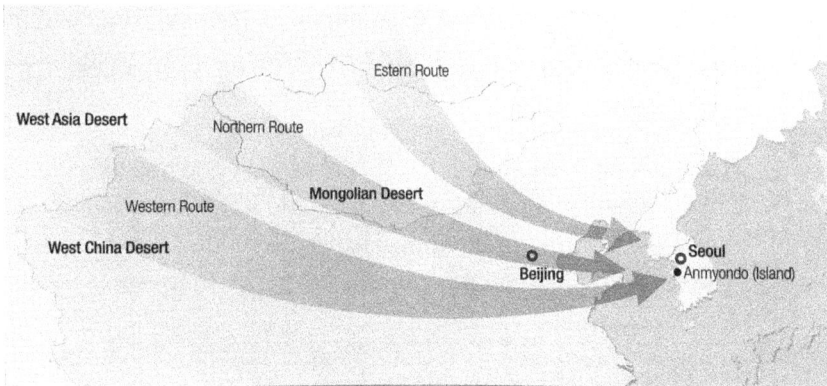

Figure 23. China, with dust aimed at Seoul. Courtesy of EcoPeace Asia.

ties between governments are at issue. It "remains something yet to come, to be desired, to be promised. . . . It belongs to the experience of waiting, of promise, or of engagement" (Derrida 1993, 368) It is a condition for a politics of friendship in the wind.[15] The dust that he and then his daughter breathe a day later, promotes but does not demand such clear delineation of positions.

Friendship begins, for the ambassador and the forest that extends his diplomatic work into the reengineering of international weather patterns, in the difficulties of history that they do not seek to escape, but rather, re-sculpt in the practical space opened at the planting site and the transnational airstream it continually evokes. Windbreaks are installed, and they are buried again. Trees are planted and they survive the season or they disappear beneath dunes they could not hold back. There are no guarantees except in the shared work of planting and relation building by future leaders, incubated in a tentative shelter that friendship, as an open-ended rapport, creates.

Forests are relations, and relations are the work itself. Not only the work of establishing and firming them and making students young diplomats that busily replicate the ambassador's work in miniature, but also, shovels in hand and everyone in the organization's logoed anoraks, they plant and dig, returning to last year's planting sites to find the business cards they tied on poplar saplings, and often having to dig a meter to find that last year's planting has been buried. A forest is composed in the wild multiplication of China and Korea down into the ground and up in the wind: here as a solemn recognition of how domestic economies have become inseparably interdigitated in Reform and Opening, here as a larger commitment to becoming part of an Asia that might shake, for a moment, long-established historical enmities and suspicions, and allows dustflows and the histories and political ecologies to kick into a regional weather pattern and shake countries into serial points in the meandering of a storm.

The ambassador's friendship in trees, then, is also an admonition, as weather everywhere becomes a geopolitical terrain: remember that on this planet, there are other ways of being earthbound to one another. The planetary need not simply reconfirm the global; the gaps in their contact might reshape them both. The engagements required of friendship are not an escape from the difficulties of politics or history. The continuity of the wind across borders cannot offer a wholesale replacement of those histories but it might allow them to phase into something else. Upwind of two capitals and a hemisphere, the friendship forest stands for the possibility of articulating a mutual protection, even while winds distribute harms unequally. The trees grow out of the understanding that, for now, the winds in a certain and not another direction will blow.

206/208, 206/207

There is a panoramic view of the Bay Area from the cafeteria of Lawrence Berkeley Labs' wall of windows. Swensen gestures over the cranes of San Francisco's building boom and over the ocean where container ships skate in and out of the bay, as if on a clear day we could see China on the horizon. By the springtime, California will be less than two weeks downwind from China, and his samples will be coated with silicates from dust storms and a cocktail of distinct geochemical traces that they drag into suspension along their course.

Isotopic analysis, he argues, will show that these traces of minerals and pollutants he gathers are Asian in origin, and Chinese in particular, especially as he has expressed skepticism over the power of computer modeling. His team has shown that in the comparative samples of air, there is a marked differential in the relative concentration of three lead isotopes—lead 206, 207, 208—which are left after the decay of uranium and thorium in the earth's surface. The ratio of leads 206/207 is a signal of uranium decay while that of 206/208 is a result of thorium decay.

The geological and radiological mega-histories that they learn to discern in isotope ratios unfold Chinese development as a question of relations through the plumes that subsume and unfold geopolitical histories and GDP into geophysical forensics. The unique geological conditions of Asian tectonic formations and their idiosyncratic concentrations of these two radioactive elements means that Chinese land has an isotopic signature that "sticks out like a sore thumb" compared to US samples. This signature holds for both mineral dusts traced to what is today China's coal soot. He argues that because the Chinese economy is powered by mostly domestic coal, and that coal powers the vast majority of the Chinese energy structure, the air can be quite confidently pinned to Chinese sources—while Chinese lab techs based in other labs in California that I speak to say that it is only possible to trace their sources to Asia more generally.

While the demands of regulators require that international borders harden again in the air's diffuse geology, the isotopes and their scientists offer another image of the Pacific. In dust streaked with the signatures of China's land and life, the China and United States that fall into sequence in the Pacific Dust Express are tethered to one another in dense and shifting bonds. Isotopic signals, amplified in scatter graphs of isotopic ratios, are the evidentiary basis by which datasets extracted from traces of dust map the hemisphere into American air. The geochemical identification of China in aerosol apportionment may eventually allow western states to skirt the Clean Air Act.

But in the China-in-the-US data set, China is more than a background to be identified to game a regulatory threshold. Atmospheric monitoring is a geological forensics. The data sets that the scientists and their devices create cannot be dismissed as ideological projections of contemporary political anxieties into transhemispheric aerosols. They are instead, an "interscalar vehicle" that reorients the universality of the Anthropocene within the ambit of concrete histories and geographies, "a means of connecting stories and scales usually held apart" (Hecht 2018, 115). The data set's scatter of points is freighted with geological memory, a month that holds the entire history of China's reform as a climatic event, that holds the dizzying scales of plate tectonics, radioactive decay, and the particulate signatures of coal, life made geological and burned into trans-Pacific airways. These airways carry an earth forming long before any place called China not to mention the combusted imprint of China's energy structures. The fluid ratio of leads 206, 207, and 208 register, in fluctuating geology, global financial markets and the succession of social-isms in the Sinocene. In the partition of airspace by the apportionment of air's contents, dustflows portend other ways of figuring relation on a changing planet.

China is upwind. Its deserts rise and fall. If thousands of miles upwind, Chinese officials are busy in their work to reengineer the dynamics of their storm-prone continent, American scientists learn to read air as an after-ness of upwind land. South Korean and Chinese students head against the route of a storm to raise forest belts upwind of capitals that are also points in the wind. In planting they create a shelter, a shadow in turbulence of land and air, where they experiment with a friendship fit to the demands of a modern weather system. Airstreams become open-air laboratories, experimental spaces that open in the tangling disjunctures of geopoli-tics and meteorology, the gaps that yawn open where world systems and weather systems touch.

"The atmosphere thickens, everywhere."

Claude Lévi-Strauss, *Tristes Tropiques* (1961, 37)

Chinese fallouts are transforming the biochemistry of the world ocean. Desert dusts are crucial to planetary ecology (Field et al. 2010), and especially oceanic iron cycles.[1] In the high-nutrient low-chlorophyll (HNLC) region in the seas off Alaska, scientists have long observed low levels of larger phytoplankton-like diatoms despite high levels of ocean nitrogen, a crucial marine nutrient for the growth of phytoplankton biomass. The growth of plankton life is inhibited by low amounts of bioavailable iron.[2] Cycles of oceanic iron were disrupted by nineteenth-century whaling, when the precipitous decline in baleen whale populations across the world ocean removed whale excrement from the upper water columns.[3] Whales, in feeding at various depths and excreting near the surface, move iron through their movement down to feed, and up to breathe. The excrement of whales was a crucial historical source of iron in the near-surface of the ocean, where photosynthetic plankton depend on it to grow (Maldonado et al. 2016). Historical whaling can be registered in the patchworks of nutrient abundance and deficiency across the ocean, and the ghosts of whales linger in the distribution of iron across the breadth and depth of the ocean.

A century and more of disruption to cycles of oceanic iron has found itself suddenly supplemented by iron-rich Chinese dustfall. Aerosol formations, like the histories of inland land degradation and its suspension into storms that pass through Chinese industrial exhausts, contribute water-soluble particulate iron along their hemispheric paths (Chuang et al. 2005). The fallouts of its modern weather, an aerosol cocktail of particle deserts and industrial emissions, ripple through the northern ocean as a ghostly echo of near-extinct whales, triggering plankton blooms where the routes of the wind and the wakes of overhunted whales tangle just below the water's sunlit surface. This amounts to a compromised nutrient reparation. Dust clouds recall the migrations of whales, up, down, and across the ocean. Acrid soot-corroded dust precipitates an unexpected reunion. The tangled wakes of disappeared whales and modern weather systems turn green with the windfall of bioavailable iron.

A cloud is, for a moment, not unlike a whale. In this accidental biomimicry, dusts and the ghostly phytoplankton resurgences in HNLC regions of the ocean become part of a menagerie of Chinese weather-monsters, whose range expands with aerodynamics of dust storms. Such monsters are unexpected entities of China's meteorological contemporary, arising in the meteorological tangle of geochemistry, ecology, and flesh.[4] Monsters, reminds Donna Haraway, define the limits of community (1991). They reveal an edge even as they undo it, gesturing at a mode of sociality and relation still to be deciphered. And monsters—sharing a root with '*demonstrate*'—*mean*. They signify, but still only in how they pose a question of relation. They signify, but in a semantics still searching for moorings in oceans of fallen land, blooming for the first time in a century. China's meteorological contemporary is rife with ecological echoes and proliferations of life in the physics of *wind-sand*.

Farther south, scientists set off into California's coastal redwood forests looking for mountain lions. Weeks after the pyrotechnic release of geology into smoke, mercury from China's coal plants settles into waters up and down the West Coast of the North American continent. Off California it mixes with mercury from the Gold Rush of the nineteenth century (Domagalski et al. 2016) released into coastal waters as a slow trickle down dammed and diverted rivers, and sometimes in torrents, with the increasingly erratic boom-and-bust schedule of rains after years of drought. This relatively benign mercury, as atmospheric chemist Peter Weiss-Penzias shows, undergoes photochemical reactions to become a neurotoxic methylated mercury. Volatile particles vaporize into gas at the surface of the water where they soak into fog and spill onto land.

Weiss-Penzias reports mercury in fog in concentrations a hundred times that of local rainwater. Absorbed by fog-loving lichens, this mercury fog is weather turned into weather, two hemispheres of heavy metal absorbed into biomass. Mercury, in its methylated form, becomes a low-level chemical constituent of a neurotoxic ecology spilling and receding along the coast, tracing the footprint of the fog through the rugged topography of coastal ranges. It soaks into the more-than-human ecology of the foggy coastal forest. While mercury concentrations remain low in lichen, they biomagnify up the food chain, first into lichen-eating deer, and finally to mountain lions, apex predators after the de-ranging of black bears. Weiss-Penzias and his collaborator, ecologist and puma specialist Chris Wilmers, measure mercury concentration in whiskers and furs retrieved from the woods, comparing this with archived mountain lion tissues and those of mountain lions from beyond the range of the fog. He pores through furs, whiskers, and claws, clipped from tranquilized pumas in the Santa Cruz Mountains or carefully archived for decades (Weiss-Penzias et al. 2019).

Life, worlded into weather, has become a permutation of an upwind geology and economy. China's meteorological contemporary, like the long-lived industrial chemicals that Michelle Murphy describes, "has joined the molecular fabric of our bodies" (2017, 495), imbricated in the enfleshment of bodies and ecosystems. It incorporates the volatility of mercury into fiber and flesh, each a site of contact with the land-enveloping fog. Ecologists and atmospheric chemists, in a haze of heavy metals, track the overlap of mountain lion ranges and the shape of the fog, compiling with their grids and sweeps into an idiosyncratic cloud atlas. In our conversations, we wonder over Californian ecologies as a stewing in the dust-spread poisons of China's rise, in a United States outsourcing production into the wind, only to have them slowly reaccumulate with each season of dust. Trade wars and toxic returns, registered in the meteorological cypher of a recovered claw or whisker.

This search for neuro-intoxicated mountain lions is a speculative zoology, a search for another weather monster. It portends China's meteorological contemporary as a plume that tangles with others, as the very substance, process, and relational medium through which planets are reconfigured. Phytoplanktons bloom in the Pacific like stepping stone islands where dusts meet the surface. The mountain lions are incorporated through fog as necrotic vessels for the "corporeal storage of heavy metals" (Graeter 2020, 22). They bio-accumulate China's meteorological contemporary and store it in self-sabotaging fatty tissues. Their altered animal cognition: a map of history, economy, and airstreams coming apart in the quiet melt of synapses.[5]

————

What world systems and weather systems, then, can be charted in turbulent phasings of earth and air? The question matters in asking how to track the environmental and political emergences in this planet hurtling past all thresholds. Aerosol entanglement and trajectories disclose China as a continent in dust, a geometeorological terrain that shifts with each twist of the wind or a dust storm broken back into land in its impact with a weather front. Worlds and experiments unfurl in this parade of permutations of earth and air, phasing from *terra firma* to dust cloud to particulate trace in a hemisphere away.

In this book, the geographies and relations that shape with dust reorient worlds to the encounter of experimental political practices with the dynamism of an unsettled Earth. Across the phases of dust, the political vocabularies and analytical habits through which scientists, bureaucrats, protestors, and others apprehend the world do not find, in the earth itself, some ready-made alternative. The earth does not offer, in these moments, a materialism that will deliver a tired humanism into its vindicated posthumanism. Rather, I have tried to show that the enterprise of thinking, intervening in, and surviving a stormy earth itself proceeds through phase shifts, where those vocabularies and habits are turned, with wind and sand, into other permutations. They have not fallen apart. Rather, they fall into other configurations, as in so many quarter-turns of a kaleidoscope.

The meteorological contemporary that this book threads through the phases of dust reveals China as a process that also splays across phases. Kuan-Hsing Chen writes that the promise of Asia as a method inheres in the possibility that "critical studies of experiences in Asia might be able to offer a new view of global history, and to pose a different set of questions"

(2010, 14–15). I stake that Asia, both in experiences and in the many ways that Asia unfolds, to upend the frame of experience when taken as an interplay of material and political emergences. Asian dusts challenge us to pose questions through the very stuff of the earth, the sky, and the very literal interchange between them. In desert, dust, and wind, China's meteorological contemporary takes shape in tangles and permutations of the meteorological and the political. This blurred jarring of domains sets loose new desires and logics of engineering and vulnerability and reconfigures old ones. In engagements with landscapes that might become weather, or awash in the dispersed solid of a dense air, institutions, meanings, and templates for political life are set on a path of remaking a more-than-human poeisis of modern land and air.

Drawn into the volatile choreography of materials that fold land and sky into storms, more-than-human assemblages of plants, waters, sands, and institutions struggle to learn how to "move and live in the whirlwind" (Berman 1988, 17). This whirlwind of the contemporary unsettles the status quos by which we understand how the Chinese state and its downwind neighbors might relate to nature and each other. Moreover, at the same time, we can witness how it draws things into its ever-changing meteorological process. The state, the unstable concepts of the political that multiply constitute it, and the political and institutional legacies of late socialist governance do not simply act on a changing landscape. Rather they become part of that landscape and change with it. Ethnography thus entails also noticing how the patterns and forms of a dynamic earth ripple through their actions and aspirations, making both the political and the meteorological sites of a continual unfolding experiment.

The question has thus not been to ask what is a politics of nature, which retains the directional sense of political action onto an inert world. Indeed, even as popular and scholarly accounts have long seen in the People's Republic of China the apotheosis of a modernist state bent on the pacification of a tamable nature, recent work has shown that the natures that haunt even official proclamations have never been monolithic. They are shot through with "variances in the very notions of nature and environment" (Liebman 2019, 2) that draw from sources as diverse as Taoism, Marxism, and new languages of sustainability. Thus, my approach has been to explore how political imaginaries, institutions, and practices are reconfigured as part of the becoming of a planet that is not only more-than-human but more-than-biological. It is "dynamic and heterogeneous, formed again and again from presences that are cultural, historical, biological, geographical, political, physical, aesthetic, and social" (Raffles 2001, 7).

The sense of the political that can form in such dynamism and heterogeneity is necessarily experimental. It is tethered, but not confined, to any preexisting orthodoxy, and eludes any simple typological grid or fixed standard of moral or ideological purity. And so, instead of looking toward environmental struggles to develop an ever-finer taxonomy of authoritarian types, I have explored how land degradation, particulate matter, and airstreams that whip across and beyond China's aerodynamic continent have spurred the flourishing of even so-called authoritarian rule as a vibrant experimental system. Late socialism thus appears as a process not only weathering the monumental environmental challenges that face China, but also as a profusion of practices wherein the Anthropocene is reconfigured into a Sinocene. It is in the experiments that chart out new permutations of state, landscape, and governance in and downwind of a country that has also become a weather system.

The wind crosses borders and dusts push concepts and institutions into phase shift. But neither is a domain of freedom. To push against the habits of a landlocked episteme is not to arc teleologically toward a liberation in the domain of the material world, no matter how vibrant it has become. And so, this book has concentrated instead in moments where form is both unstable and emergent, where dissolutions of one logic foment new attempts at capture by another. This book has thus cataloged what happens at the interfaces of domains that might otherwise be considered separately: late socialist statecraft and desert winds; sand-seas and the dust-crossed Pacific; diplomatic and meteorological ties in a complicated Asia; the mobile landscapes and ambient thicknesses that phase with each other to make modern weather.

This is necessary because I do not wish that the anthropological maneuvers this book has worked through be taken as necessarily progressive. Instead, I see it as populated with strange moments of para-ethnographic parallax, in which the tenets of an ethical anthropology and an experimental ethnography have already been implicitly adopted by, among others, the Chinese Communist Party in its local operations.

The scientists, bureaucrats, and engineers that struggled to reshape social worlds and the textures of their earth to blunt the power of the wind knew to take seriously a more-than-human perspective, precisely as they seek to reengineer those relations. They nonetheless also teach us how to see like a state in the attunement to the shape of a dune. They taught me to see desert botany for its multiple potentials, even as they might structure bureaucratic experiments that would move populations, human and otherwise, and recircuit late socialist institutions in order to control the movement of expanding deserts. The ethnographic demand to study from the specificity of place found its grotesquely fascinating funhouse doppelgänger in CCP theory in which the already suspect Anthropocene contorts into a Sinocene.

But sometimes, wading in dunes that shift to engulf ankles, or being twisted into the wind's twist, there is a glimmer of something else. The possibility of a friendship in this strange weather; or, the sense of shared fate across hemispheres, extrapolated from a trace of earth carried two weeks on the spring wind. A shared, coughing wonder in peals of dust and the shapes that unfurl with it. The wind will not cease. Experiments are open ended and so are planets. Neither is easily assimilated to narratives that require endings.

Notes

ACKNOWLEDGMENTS

1. From "Starting with a Line by Joyce Byers" by Eric Tran (2019, 3, lines 21–25).

APPARATUS A. NIGHTWIND

1. This and all other names in this text are pseudonyms.

2. The first five-year deadline for the pilot resettlement program to determine the status of these houses, as well as government-provided earthen greenhouses (*wenpeng*), was slated for 2013, after the main period of fieldwork. The decision has been deferred indefinitely. As of follow-up research in the summer of 2017, a decision had not been made, leaving the Tais and other ex-herder families in the resettlement village in a state of legal limbo. This precarity is discussed in chapter 1 in the section "Autopilot."

3. Much of the sometimes-circuitous writing in this book is an attempt to consider how in "a lively, more-than-human world" language must contort to pose other modes of agency than that of "the willful individual," always human, "positioned above the fray" rather than in it. Jane Bennett poses this pursuit through the necessary use of "middle-voiced" verbs that "bespeak from *within* an ongoing process, rather than from an external vantage where the subject of a

predicate can either direct activity (active voice) or be acted upon (passive voice)" (2020, xvii–xviii). Please read this book with my apology in mind, as I am after a language that does not require an agent that acts or a subject that is acted upon. Instead, let us envision what Bennett calls a middle voice, present in some non-English languages, that figures happening in a field of activities and processes in a world in influx and efflux.

INTRODUCTION. EARTHLY INTERPHASES

1. The seasonal change in wind patterns is also known as the monsoon, usually most associated with the Indian Ocean. The monsoon has, in several iterations of area studies, been understood as a meteorological-cultural basis for understanding the coherence of Asia as a unit of study, under the rubric of "monsoon Asia." While long understood as a durable and predictable feature of the region's physical geography, the monsoon appears to be changing. Some of the cultural and political effects and security risks associated with this change are documented in Nidhi Mahajan's historical ethnography of dhow seafarers (2019) and Sunil Amrith's environmental history of Asia through water (2018).

2. Atmospheric scientist Daniel Jaffe's group's airborne research has been crucial to my understanding of dust modeling and visualization.

3. The morphology of plumes offers a way of thinking about mixture without the even distribution of solutions, and requires a material analytics focused on interaction and fluid dynamics. Andrea Ballestero's writings on plumes in "hydrolithic choreographies" of underground aquifers rhymes with this book's formal and material sensibility by "moving away from our implicit comfort with thingness and closer to the intermittence of process, to a material-semiotic history that has always been jumpy and unpredictable" (2019b).

4. Dust events take different forms and operate at different scales, depending on factors like variations in particle size, airspeed, and visibility. Goudie and Middleton (1992) offer a schematic typology to define dust events driven by aeolian processes: dust storms; dust haze; blowing dust; and dust devils or dust whirls.

5. Corey Byrnes (communication with author, March 21, 2020) delightfully points out that the image "looks like a Chinese landscape stone, or *shihua*, sometimes called a dream stone in English. These are ornamental stones, often set in frames or furniture, with patterns that resemble landscapes."

6. Hugh Raffles describes other scenes of full sky in his entomological inquiries. He reminds us that the "deep vastness of the air, the heavens wide above you" are "full of insects, and all of them are going somewhere" (2011, 12).

7. Reluctantly paraphrasing William Blake (1950), what worlds do grains of sand make?

8. "Reform and Opening" refers to a period of liberalizing reforms made by the CCP under the leadership of Deng Xiaoping. During this period, agriculture was de-collectivized, foreign investment was allowed, and the private sector expanded as worker entitlements (like the Mao-era "iron rice bowl") were steadily rolled back.

9. I am inspired by Aimee Cox's sense of choreography as "shapeshifting made visible" (2015, 28), an apt description of the political and material choreographies that I gloss in this text as the weather.

10. Throughout this text I insist on *wind-sand*, an italicized calque, and not its closest English translations, blown sand or blowing sands. I make this decision partially to preserve the grammatical openness of the Chinese compound word. Only in translation does the question of specifying how wind and sand relate become necessary. As consecutive nouns? As an adjective and the word it modifies? My usage here follows translator Ken Liu's precept: "The best translations into English do not, in fact, read as if they were originally written in English. The English words are arranged in such a way that the reader sees a glimpse of another culture's patterns of thinking, hears an echo of another language's rhythms and cadences, and feels a tremor of another people's gestures and movements" (2014, 398). The point is to speak to the generative displacement of movement across unassimilable domains. Translation's ethnographic work is not then in its fitting of seamless correspondences, but in its "talent to maintain divergences among perspectives proposed from worlds partially connected in communication" (de la Cadena 2015, 27). This is, of course, a crucial device in anthropological thought.

11. I glean the metaphor of folding from Aihwa Ong's discussion of postgenomic science in Singapore. Origami, she suggests, refers "to the folding of disparate systems of code" into unexpected connections. "Origami-like relations" (2016, 16), made and made again in folding, offer something close to the universe of possible patterns that I seek to evoke with the figure of the kaleidoscope. Consider also Hao Jingfang's science fiction short story "Folding Beijing" (2014), which figures the city as a fractured populace, segregated into layers of the city that take turns being unfolded in underground layers and into the surface.

12. The sense of blown sand as a composite substance that is also an array of possible formats of sand and wind is fundamental in the disciplines of aeolian physics. As they pose their object specifically in an at least dyadic relation—the hyphen in *wind-sand*—these modes of attending to substance, process, and relation offer important ways of picturing how, virtually and actually, the "context of contact" (Howe 2019, 11) that wind formats environmental transitions and emergences. This sense of substance as transformation is crucial. It invokes what Deleuze and Guattari describes as the collapse of a distinction between substances and forms. "Substances," they write, "are nothing other than formed

matters" (2000 [1980], 41). In *wind-sand*, substance is less a question of being formed than a question of the possible forms that a relationship might take.

13. In his writings on the Three Gorges of the Yangtze River, Corey Byrnes shows that landscapes are not unproblematic givens. They emerge in the "interaction of the representational and the physical" (2018, 5), their futures shaped by poetry as much as wind, and technology as much as water. To extend his thinking from landscapes to weather systems is to approach weather as a zone of "ideological and material encounters with traces of China's past that opened up new imaginings for China's present and future" (Karl 2020, 4).

14. The imbrication of dust into bodies is reminiscent of Michelle Murphy's thinking on environmental toxicology and late industrialism (cf Fortun 2012). Murphy calls this relational mode of bodily becoming "alterlife," "life already altered, which is also life open to alteration" (Murphy 2017, 497). Alterlife anticipates Stefanie Graeter's notion of "mineral incorporations," attentive to how materialities imbricate bodies, futures, and ecologies into toxic emergence (Graeter 2020).

15. Stephen Collier proposes topologies of power as "patterns of correlation and recombinations," with an attention to "how existing elements are taken up and recombined" (2009, 87, 90). The notion of a topology is generative for me because it allows us to attend to specific strategic mobilizations of governmental practices, without the typological and periodizing impulse to associate fixed ensembles of governmental practices and logics with specific stages or historical periods.

16. Dust events have long been attested in Chinese and downwind historical sources (J. Zhang and Crowley 1989, Masatoshi 2002).

17. One pleasurable disorientation of this writing is in how the attention to wind through fluid dynamics gathers wind and water as analogous fluids. Watery vocabularies are thus transmuted into a general language of fluids, of liquid or air. "The mechanical fundamentals of aeolian sediment transport," writes geologist Michael Welland, "are the same in many ways as those for the actions of water in rivers and oceans" (2009, 151).

18. The question of the relationship between China and its land has been a contentious one. Grace Yen Shen notes, in her history of geological sciences in China, that "attention to the materiality of the land" (2014) was shaped by contentious questions over what constituted Chinese-ness amid momentous social and political changes. Fei Xiaotong argues that Chinese-ness—or at least the majority Han version of it—is inextricably rooted in the very substance of the soil, which migrants secret with them as a physical metonym of place, kinship, and self (1992).

19. Earlier geographies of national insecurity, also routed through Beijing, offered an affective and administrative scaffold that the wind would both invoke and reshape. Beijing was designated in 1421 as capital of the Ming Dynasty as a

check to the expansion of non-Han invaders. They crossed the Great Wall from regions long associated with incoming danger.

20. As Anna Tsing writes of places constituted as remote, central power in marginal places is largely an imaginative enterprise, rife with interpretations and misinterpretations rather than the straightforward implementation of force. She emphasizes the ways in which marginal and central places coproduce and also mis-produce one another. The "imaginative features of nation-building, ethnic formation, and state rule as experienced from the margin" (1993, 287) matter powerfully in the introduction of meteorological proximity as a supplement to remoteness as a key feature of Alxa's "locality."

21. In premodern China, imperial regimes were subject to a "moral meteorology" by which weather anomalies could be mobilized as divine commentary on the righteousness of rulers (Elvin 1998, 213). The resonances of this moral meteorology play a part in conditioning political experiments for controlling *wind-sand*.

22. For excellent overviews of the history of desertification as a concept, see D. Davis (2016, 2017). The term *desertification* was introduced in 1927 by the French colonial forester Louis Lavauden and was closely associated "with notions of spreading deserts and of human culpability for deforestation, overgrazing, and land degradation" (D. Davis 2017, 1). At the center of desertification debates are questions of how political economic regimes both drive landscape change and, also, how ecological processes in non-equilibrium environments come to be sites for the powerful reorganization of governance.

23. Please note that emergence is not a figure of existence out of nothing, or novelty without history. Following Paul Rabinow, I think of it as "a state in which multiple elements combine to produce an assemblage, whose significance cannot be reduced to prior elements and relations" (Rabinow and Bennett 2008, 398). The language of de- and reterritoralizations (Deleuze and Guattari 1980) is close to what I mean by phase shift. I do not invoke that language here, however, because I want to retain the crucial sense that the earths implied in territorialization do not come undone as much as they continuously reconfigure in multiple terrestrial permutations.

24. The question is resonant with Cymene Howe's anthropology of aeolian politics on the Isthmus of Tehuantepec. Her anthropology "defamiliariz[es] . . . 'wind power' as a singular, technical, managed form" (2019, 7) and attends to its force, its indexical relationality, and its tethering of possible, divergent futures.

25. Michel Serres writes that in a disoriented technological present there is a confusion of weather and time: "We surely don't know how to think about the relations between time and weather, *temps* and *temps*: a single French word for two seemingly disparate realities" (1995, 27).

26. The contemporary is also not synonymous, in any straightforward way with "the present" if the linear and singular timeline of modernist calendars are

assumed. It is instead a way of asking how times fold together (see Arendt 2006). Michel Serres gets at this when he considers the nature of the contemporary through the example of a late-model car. "It is a disparate aggregate of scientific and technical solutions dating from different periods. . . . The ensemble is only contemporary by assemblage, by its design" (Serres and Latour 1995, 45). Its novelty is in how specific pieces, how disparate processes and irreconcilable histories, come together.

27. Rather than endpoints, we think through moments in which the capacity to "become self-different" (de la Cadena 2015, 280, cited in D'Avella 2019, 21) is realized. The resignification of fixed categories into processes of transition is a conceptual and metaphysical move that, for me, has many precursors, for instance in classical philosophies, Chinese and otherwise, that attend to matter through its transformation. Elements are not irreducible stabilities, but capacities to change.

28. In posing the question this way, I invoke Candis Callison's question, "how does climate change come to matter" (2014) in inverse, emphasizing the topological possibilities of syntax for conjuring relations. In this inversion the social sciences and meteorology appear as permutations of one another, recalling various anthropogeographies that, in other ways, explicitly pose weather and human sociocultural phenomena together (see Bunzl 1993, 25–27).

29. I am indebted to a vibrant conversation in anthropology, geography, and the humanities (McCormack 2018, Adey 2014, Flikke 2016) that has insisted on the atmosphere as a correction and alternative to landlocked thinking. Cymene Howe, in her introduction to a special section of *Cultural Anthropology* calls for an exploration of atmosphere as essentially a "deterrestrialization of thought," that explores "vitalities, materials, and movements that are skyward, spacey, and atmospheric" (2015, 203). At times, this work may be read as suggesting that air is an alternative or escape from land toward "non-earthly places" (2015, 204), which suggests that air and earth are analogs of freedom and constraint. I emphasize that the relation between air and earth should not be one of binary opposition but of substantial entanglement. This builds on Howe's point, in the same essay, that attention to life above earth "draws our attention to life *on* earth" (2015, 206, emphasis in original).

30. See Paul Edwards's *A Vast Machine* on the political and technological history of meteorological informatics (2013), Joseph Masco on the imbrication of earth systems sciences with Cold War nuclear programs (2010), and James Fleming's (2012) and David Biggs's (2018) histories of military meteorologies. Michelle Murphy's work on engineered indoor airspaces (2006) and Peter Sloterdijk's writings on atmoterror and gas warfare (2009) emphasize the capitalist and military histories of meteorological intervention at various scales.

31. In a discussion of Chinese thought and aesthetics, François Jullien considers blandness (*dan*) as an aesthetic quality of minor differentiation and

indeterminacy that invites reflection and the sharpening of appreciation of subtle variations and inconclusive sensory and semantic experience. That which is decidedly bland, he writes, "can give rise to the richest variations and farthest-reaching applications. Now meaning can never be conceived as closed and fixed but remains open and accessible" (2004, 33). The blandness of "weather" allows for open-ended tracing of meaning, which the schema of pollution, with its ready-made positions (polluter) and sense of causation *cum* blame, cannot accommodate.

32. William Cronon critiques the idea of wilderness, cautioning against "any way of looking at nature that encourages us to believe we are separate from nature" (1996, 22). The idea of wilderness, for Cronon, makes it impossible to attend to more-than-human relation. This insight, that "pollution" is an ironic mirror of Cronon's notion of wilderness, is Stefanie Graeter's (personal communication October 25, 2020).

33. Timothy Ingold describes "weather-worlds" as a zone of meteorological and material fluxes in which all life is lived. In weather-worlds "we do not so much traverse the exterior surface of the world as negotiate a way through a zone of admixture and interchange between the more or less solid substances of the earth and the volatile medium of air" (2010, S122). The notion of a weather-world offers a way of considering, anthropologically, the relations between ground, body, being, and air as interchanges in weather. Weather-worlding brings this conversation productively in relation to "worlding projects," "emergent, transformative relations and processes deeply and inexorably enmeshed in sociohistorically contingent productions of difference" (Zhan 2009, 7).

34. Amitav Ghosh (2017) writes subtly on derangement as a means of capturing the multiple entangled disorientations that drive climate change while also foreclosing the possibility of addressing it. For Ghosh, derangement is at once a quality of climate systems thrown into unpredictable and chaotic variation as well as the abdication of modern literature's responsibility to provide the imaginative resources to consider that variation.

35. For scholars of the USSR, late socialism has generally been a periodizing gesture, referring to the decades 1960–1990 (Yurchak 1997, Kaganovsky 2009). I insist on late socialism as a way of holding in suspension the irresistible momentum of narratives that have claimed the substantive end in socialism in China and its historical reconciliation into neoliberalism (Harvey 2011). Post-Soviet ethnographies of the former USSR and Eastern Bloc have powerfully cautioned against the romance of these transitionologies. Analytics of transition are based in two reassurances, offered as mantras to exorcise the specter of communist returns (Derrida 2006): that socialism has ended, definitively consigned to the trash heap of history; and that the future can be imagined as, above all, in the teleological movement toward capitalism, liberalism, and democracy. Posing late socialism through the mutual traction of environmental and political

change, I follow Simondon's philosophical notion of the individual, which "can be grasped as an activity of relation, not as a term of this relation" (2020, 50).

36. I follow Fredric Jameson, who notes that the "late" in late capitalism serves "to mark [a] continuity with what preceded it" (1992, xix), rather than indulging in a periodization of hard breaks and ruptures, which require, of course, endings and beginnings.

37. This way of thinking East Asian politics is clearly deeply indebted to Aihwa Ong (2006). I depart from critiques of modernist and megalomaniacal pretentions of human control over an external nature. This kind of critique constitutes "nature" as inherently resistant to its pacification by dint of its own vibrancy, or the limitations, usually anthropocentric, of whatever political formation is in question. This geometry of power, which requires a state that acts and a nature that is acted upon, is well critiqued as a high modernist fever dream, proceeding through the simplification of ecologies in accordance with the short-term dictates of markets and rule (Scott 1998, Tsing 2015, Haraway 2016). It finds its Chinese analog in accounts of the anti-scientism and hubris of "Mao's war against nature" (Shapiro 2001), and its forceful, quasi-military imposition of an ideological hard line on a universe of vibrant things, each of which mounts, ontologically, a small war of resistance.

38. Late socialism is, in this sense, a rejection of the term *postsocialism* as a periodizing concept meant to convey that socialism is over. (Although the concept originates in scholarship on the Eastern Bloc and former USSR, it is sometimes used to diagnose a Chinese contemporary). The problem with post-socialism is that it implies socialism is a historical period rather than a complex structure of continuously changing logics, modalities, and styles of political governance (Collier 2011, Yurchak 2005). It also completely dismisses the possibility that socialism could be a possible future and a way of composing that future (Derrida 1993), or that it could be an ethical stance—a "sign-value" that "call[s] to responsibility" (Anagnost 2004, 206).

39. But if experimental systems are definitionally dynamic and pluripotent, they still tend to be "described by highlighting their fixity, as if [their] components were already predetermined" (Ballestero 2019a, 9). From this vantage point, such systems appear to be locked in cycles of repetition that arc toward failure and obsolescence.

40. Deng Xiaoping describes what in retrospect appears as preordained "transition" to a marketized economy, in terms remarkably familiar to anthropologists that revel in bricolage. Reform is a practice of "groping for stones while crossing the river." It is, in Ezra Vogel's description, "a creative way of encouraging experimentation and acknowledging that in a new situation [leaders] should not expect that all policies would work well" (2013, 391).

41. In her exploration of desire as a crucial site where late socialist power operates, she shows how the instability of desire itself does not guarantee any

particular outcome or political form, but constitutes "the social field of desire as experimental" (2007, 23). For Rofel, desire is not only an emergent concern for the reconfiguration of late socialist power, but also an excess that continuously forces that power to shapeshift in tandem with its instability.

42. Yifei Li and Judith Shapiro helpfully show that the official principle of "ecological civilization" enshrines green initiatives into the history of China under the Chinese Communist Party as a continuation of the legitimacy of rule as well as a new phase in the official historiography of national progress. From agricultural to industrial to revolutionary and finally to ecological civilization under Xi Jinping, this progressive stagist model is "a faithful reincarnation of Marx's theory of the stages of development with Chinese characteristics" (2020, 6).

43. The militarization of nature as an enemy to be conquered presages what Frédéric Neyrat describes as "the fundamental fantasy at the heart of geo-constructivism," a set of discourses that figures the Earth in its ultimate controllability. This fantasy claims that "the Earth, and everything contained on it—ecosystems and organisms—humans and nonhumans *can* and *must* be reconstructed and entirely remade" (2019, 2).

44. Savannah Shange notes, "Built into the genre of ethnography is an expectation of narrative thickness, a rich tapestry of voices that leaves the reader satiated by the elegant rhythm of *I saw, she said, I saw, she said*" (2018, 119). Shange reflects on the refusal of Tarika to yield to ethnographic insight as a point to reflect on how such intersubjective encounters are aestheticized as part of the formal literary entailment of ethnography *per se*. Here, I depart, lovingly, from Shange's intervention, not to probe the ethics and politics of refusal, but to raise a methodological question for an ethnography in which human life matters, deeply, but as part of and in relation to a weather process that upends the generic requirements of ethnography when imagined solely as a study of human life.

45. Haraway's account insists on the primacy of vision despite its status as "a much-maligned sensory system in feminist discourse" (1988, 581). The difficulties with the familiar example of human (or primate) eyesight is, of course, its naturalization of vision as unproblematically embodied, and, relatedly, the ableism of holding that mode of embodiment as a general principle of knowing. Hereafter, I refrain from identifying parallax with human bodies. After all, not all humans sense in parallax, and not only humans (and primates) sense through displacement.

46. Extraterrestrial perspective might evoke lofty visions of a planetary commons, and lay the imaginative conditions through which new social, technomilitaristic, and environmental connections can be phrased (Lazier 2011); or, here, in their displacement of the habits, they disorient how one seems already to know what a country, a space, an earth are.

47. This multiperspectivism has been a crucial part of a much longer argument about ethnography and reading more generally. Eduardo Viveiros de Castro's

accounts of Amazonian perspectivism also put forth "an ethnographically-based reshuffling of our conceptual schemes" (2012, 46) through the "human-animal parallelism" (2012, 50) that analogically territorializes subjectivity in multiple human and more-than-human perspectives.

48. It is perhaps also worth noting that parallax offers an idiom and a physics for thinking through the un- and ir-resolved dynamics of the author's own ethnographic location, which would mean that situating oneself is less about proposing identity positions that one occupies than pointing to the systems of differences among which that occupation is rendered impossible. This may seem coy. But my proposition of parallax here is an attempt at posing the space in "Chinese American" ethnographer not as an intermediary or transitional zone, but as a wild field of oscillations.

49. Migrating birds, for instance, fall into flying-V formation. The dynamic pattern of birds in wind is a realization of the shape of their relation—they cannot be abstracted out of the medium that holds them together. The ensuing aerial form "unites the individual birds" (Schwenk 1965, 116) in a variegated aerodynamic totality that exceeds any of its constituent parts. Jason Price turned me on to Schwenk's work when he, anticipating my excitement, borrowed the book from the UC Berkeley library and gave it to me. I proceeded to spill coffee on it, and then had the gall to give it back to him to return to the library as if nothing had happened.

50. Marilyn Strathern quips, "Scale has been a headache for anthropology" (1995, 15)! She is right!

51. Readers may note a recurrent analytic form in the book's two halves. Chapters 1 and 4 describe a specific format of relation through *wind-sand*. Chapters 2 and 5 elaborate it as it ramifies through a profusion of experimental forms. And chapters 3 and 6 witness it through the breakdowns and decompositions that presage a phase shift, when one pattern's dissolution is the condition for another's form. Patterns coalesce, involute, and shift into still more patterns. The dispersion of social and geophysical engineering programs into scenes of maintenance, just-in-time aversions of breakdown, and erosion, in the first half of the book, become the material and political preconditions for the particulate emergences that matter in the second. One pattern frays into another and kaleidoscopes keep turning. Shape, shaping, reshaping.

52. Hong (2015) also notes that the history of desert research in China is closely tied to aeolian physics and a concentration on aeolian, or wind-driven, desertification as the principal mechanism of semiarid land degradation. By 1999, desertification research had become a mandate of the State Forestry Administration. This suggests a reformatting of forestry as a techno-ecological practice of manipulating aeolian physics.

53. This line in its original form is prose from the digital project *Feral Atlas*. The original line: "The notion that the cause of death associated with the 'fog' is natural is contradicted by the smell, the taste, the burns to the throat, the itchy

tongue; . . . by the feeling of suffocation, the heaviness of breathing . . . denied by the dust and industrial gases that *those who live in the valley are continually forced to breathe*" (Zimmer 2020, emphasis added).

54. Isabelle Stengers writes trenchantly that scientific and other practices involve "scientific passions" (2010, 1). These entail practical and ethical commitments to knowing beyond oneself through theorems, implements, and commitments that call into question the habits and limits of a self-contained knowing subject. I am after a mode of concrete thinking that is less about the confirmation or cumulative increase of an already-given style of thinking, and more about the awkward and generative process of conceptual becoming that she describes.

APPARATUS B. THE WIND TUNNEL

1. The standard translation of aeolian physics in Chinese is *fengsha wuli*, the physics of *wind-sand*.

2. For a description of the tunnel, see Zou et al. 2001. A very early description of the wind tunnel apparatus appeared in *Cultural Anthropology* in a paper cowritten with Tim Choy (Choy & Zee 2015).

3. Decades of tunnel experiments, in China and abroad, have fixated on blowing sand to picture sand, sediment, and the earth itself insistently through fluid dynamics. This wind tunnel is a distant descendant of the experimental machineries built by British army engineer R. A. Bagnold, inspired by his fascination with sand and dunes in his interwar posting in Libya. In additional to field tests, Bagnold tested collected Libyan sands in a purpose-built tunnel apparatus in Imperial College in London, centering an imperial geography in blown sand squarely in the metropole. For more, see Bagnold 1941, 2010 and Zee 2015.

4. Claude Lévi-Strauss, in *The Savage Mind* (2000), describes reduction of scale both in terms of sheer size and in complexity ("number of properties") as a means of overcoming a resistance to understanding inherent in things themselves. Reduction, he muses, is a condition of knowledge that works by allowing access to a thing as a reduced totality rather than as a complex whole. For Lévi-Strauss, all modeling must be understood through a prior act of reduction, by which an overwhelming array of parts appears instead as a small whole, visible all at once: "In other words, the intrinsic value of a small-scale model is that it compensates for the renunciation of sensible dimensions by the acquisition of intelligible dimensions" (1961, 16–17). For another account of horizoning work in projections of environmental time, see Petryna 2018.

5. In Hacking's usage, the phrase *historical ontology* draws attention to how things are constituted within specific historical and epistemic circumstances. He insists that tracking distinct "comings into being" (Hacking 2002, 5) is crucial to understanding how intervention can be imagination.

6. Recursively, the interest in land as definitively and dangerously mobile in its intra-action with air and water recalls Mao-era mass mobilizations against erosion in mountain terraces. The social *and* physical transformation of the erosion-prone countryside was at the center of the revolutionary slogan "Learn from Dazhai," a village that was "promoted as a model of communal living" (Zhao and Woudstra 2007, 171), not least because of anti-erosion social and environmental engineering. See also the following chapter.

CHAPTER 1. MACHINE SKY

The epigraph is Ursula Le Guin's personal commentary, from a footnote in her translation of the *Tao Te Ching* (2019), attributed to Lao Tzu. Chapter 28, "Turning Back," is a play of paradoxes and reversals. Especially relevant to this chapter is the paradoxical passage: "Knowing light / and staying dark, / be a pattern to the world. / Being the world's pattern / of eternal unerring power / is to go back again to boundlessness." For more on Ursula Le Guin's expansive and generous work as a site for speculation on anthropology itself, please see Pandian (2018; 2019, 97–103), Shange (2019, 12) and Povinelli (2008).
    1. I borrow the notion of an engineered world from the Engineered Worlds project, spearheaded by Joseph Masco, Michelle Murphy, Tim Choy, and Jake Kosek.
    2. By framing landscape interventions in Alxa as part of a dispersed political apparatus of meteorological engineering, I am following ethnographic interlocutors who insist on apprehending meteorological phenomena across long distances not simply as natural disasters that the state must prepare for or simply to respond to after the fact. "Among [natural disasters'] social effects is to bring infrastructure suddenly and painfully to our awareness," writes Paul Edwards (2003, 193). One other social effect, more paradoxically, is that they may seem to generate the sense that everyday life and environment are not infrastructuralized enough. Edwards in the same essay argues that infrastructures both make and *are* environmental, in the sense that they are the substrate for modern life. This suggests that not only might infrastructures be evaluated in terms of their working or failure, but that disasters might prompt shifts in the conceptualization of phenomena or conditions usually taken to be environmental as problems of an infrastructural nature—in multiple senses of the word *nature*.
    3. In his writings on the Three Gorges, Corey Byrnes shows that landscapes are not unproblematic givens. They emerge in the "interaction of the representational and the physical" (2018, 5), their futures shaped by poetry as much as wind, and technology as much as water. In doing so, he reminds us that earth and air are a technopoetic process, where there is a recursive and generative relationship between aesthetic traditions, technical practices and fantasies, and the physical affordances of environmental configurations.

4. Consider how these terraforming, earth-working projects both invoke and depart from James Scott's discussions of modernist schemes to dominate, order, and rationalize nature, especially in his account of "scientific forestry's project of creating a simplified and legible forest" (2008, 30).

5. Wading into the semantic multiplicity of "terraforming," Karen Pinkus and Derek Woods note that it offers a "constitutive role for the fantastic and the fictional in a time when any cultural or political event can be read in terms of its possible effect on the earth system" (2019, 4). Woods notes that terraforming both undermines and reinvests in the specificity of the Earth through a notion of "earthlikeness" as both changing and provisionally changeable. In the midst of climate change, terraforming has become "a means of understanding both what some humans *have* done to the earth system and what some *might* do, going forward" (2019, 7).

6. I glean the language of programming from Vilém Flusser (2013), which he distinguishes from finalistic and causal "images," exemplified by doctrines of predestination and a modernist scientific ontology of determinist, and sometimes chaotic, accounts of happening. Both, for Flusser, expunge the possibility of chance and freedom. By programming, he diagnoses modes of thinking that include figures of both constraint and possibility. Dynamic starting conditions, in a program, can and will lead to unexpected outcomes, and "chance becomes necessity" (22). This is a way of speaking to open-ended nondeterministic accounts of process, while at the same time retaining both a sense of formal parameters and their necessary malleability—conditions crucial for environmental engineers in the situation I describe. There is a longer international history of programs for modifying weather and climate, at an array of scales, and enacting different conceptions of environment, population, and power, from colonialism to contemporary debates over geoengineering the planetary biosphere to remediate climate change (Long et al. 2011). Fantasies of local climate modification, for instance, featured in colonial projects of all different kinds, as in French colonial efforts in the Maghreb to "resurrect" the region to its alleged pre-Arab status as the "granary of Rome" (Davis 2007). Nineteenth- and twentieth-century US efforts at weather modification in cloud seeding, especially in order to secure tactical military advantages, under the presumption that weather manipulation could be deployed as a strategic force multiplier (Fleming 2012, 170–74). In modern China, cloud seeding has been deployed as an ad hoc technique for clearing pollution, most famously when the Beijing Meteorological Bureau, "under increasing pressure to give its best performance yet" (Qiu and Cressley 2008, 974) used weather medication technologies to induce good weather for the 2008 Beijing Olympics.

7. China has a high ratio of engineers as political leaders. This overrepresentation of engineers has historical roots in the Communist Party and in the purges of the Cultural Revolution. Susan Greenhalgh in her work on the central role of a coterie of rocketry engineers as proponents and architects of the One Child Policy (2008) importantly argues that engineers and the configuration of social

intervention as an engineering problem have had an outsized influence in shaping social policies.

8. Approaching antidust projects as earthworks (cf. Gladstone's 2019 discussion of Michael Heizer and Robert Smithson) helps us move away from an account of infrastructure that refers only to the imposition of human design onto a neutralized nature, or to the charismatic hardwares whose triumph or breakdown has stood for promises of modernity and their collapse (Larkin 2013; Anand, Gupta, and Appel 2018). These imaginaries may make it difficult to note other ways that infrastructures might operate—for instance, in the poetics and politics of disrepair (Chu 2014, Schwenkel 2015) or in the lifeworlds and sensoria that develop in conditions of breakdown (Larkin 2008, Hirschkind 2009). Instead of asking whether an infrastructure works—and celebrating breakdown as an upset of power and its promises—we might instead attend to how they function, especially when they do not "work" in a normative sense.

9. See Wu, Wang, and Chi 2013.

10. Among excellent work on the history and political ecology of erosion, see S. Freeman 2017, Blaikie and Brookfield 2015, Worster 2004. China's Sloping Land Conversion Program was implemented after major mudslides in 1997 to prevent erosion on sloping land. See Xu et al. 2006.

11. Critiques of greenwashing conceive of it as an ecological apology to and obfuscation of extractive logics. See also works on ecologies beyond greening (Cohen 2016).

12. The notion of sand commodities or sand products (*sha chanye*) was proposed by the revered Chinese rocket scientist Qian Xuesen (sometimes Hsue-Shen Tsien) (Qian 2012). Qian was educated in the United States but deported during the Red Scare of the 1950s back to China. A towering figure in China's rocket science and space programs, he posed the notion of sand commodities, the promotion of sand-loving agricultural products, in his theorization and recommendations for steering Reform Era China toward a so-called Sixth Industrial Revolution.

13. This continuing registration effectively means that they are living, as villages, in a kind of exile, away from their lands. Many residents are hesitant to relinquish their household registration on pasturelands that they have either left entirely or only return to seasonally, because certain government benefits, like forestry subsidies disbursed for *not* grazing land, are calculated on an area basis. Formal registration to a large piece of land means that exile is also a way of maintaining cashflow through anti-sand forestry programs that incentivize having large pieces of unused land.

APPARATUS C. A SHEET OF LOOSE SAND

1. Pheng Cheah identifies a recurrent "analogy between organic life-forms and the technic of social and political organization" (2003: 230) as a prominent

feature of the theory of the nation from European Romanticism to postcolonial liberation movements. The nation, in such a view, is not merely a body, but an organism: a living, organized thing, animated by a principle analogous to the life force that sustains every other living thing. National and political survival here is to be read literally, as a *sur-vie*, a living-on of a political entity identified and endowed with life, and thus in danger of sickness, of death. The constitution of the state as such unfurls in analogy to a living body that holds its constituent parts, its organs, in relationship to each other.

CHAPTER 2. GROUNDWORK

1. An earlier and substantially different version of some of this material was published in the edited volume *Frontier Assemblages: The Emergent Politics of Resource Frontiers in Asia*, edited by the incomparable Jason Cons and Michael Eilenberg (Zee 2019).

2. "Banner" (*qi*) is an administrative unit in Inner Mongolia.

3. *Tuimu Huanlin* has sometimes been translated into English as "Grain for Green," emphasizing money transfer as its main policy feature. I follow geographer Emily Yeh (2009) and others in translating the program as Convert Pastures to Forests, a much more literal translation of the title, in order to emphasize what the program itself emphasizes: land use and landscape conversion, bolstered through payout mechanisms of different kinds.

4. A *mu* is a Chinese unit for the measurement of land area. One mu measures 0.165 acres. It is used officially and is the default unit for describing the sizes of pastoral and agricultural plots by local officials and farmers and herders alike.

5. Desert-broomrape was first described in 1960 for the US National Plant Germplasm System in the *Acta Scientiarum Universitatis Intramongolicae* (see also USDA 2020).

6. This magic, for Frazer, is an account of efficacy in similar forms, such that things that resemble one another are held to share essence. The explicit phallic shape of the root is lost on very few and is often the basis of bawdy jokes. In addition to virility, rou congrong boosters in Alxa hawk its pharmacological applications in a wide range of health domains: tonifying organs, balancing internal bodily *qi* and heat, and improving everything from digestion to immunity and cognitive function. I was advised not to drink *rou congrong* teas in the summertime. It was also recommended to me as an excellent gift for older women.

7. *Suosuo*, the saxoul shrub, as many area forestry professional explained to me, has been introduced as a preferred species for afforestation as a corrective to earlier dalliances with more charismatic species like poplar and pine. Planted widely, first as shelterbelts to mark the edges of sedentary settlements and cropland (*suobian lin*), the widespread promotion of these earlier species is widely lampooned for its political and spectacular, rather than functional, efficacy. The

quickly growing trees *look* like forests, especially compared to fields of shrubs, and so made good photo ops, standing dramatically in lines cut against stark yellow sand. They were described to me, by forestry officials, as unfortunate evapotranspiration machines, whose deep, water-seeking roots and wide, flat leaves functioned essentially as valves for channeling precious desert groundwater uselessly into the air, thus exacerbating the land degradation they were meant to contain, while dying after a few photogenic years.

8. Plants and politics relate in "simultaneous connection and divergence" (Blaser and de la Cadena 2017, 189) of disparate logics, futures, and species: the "environment" of economic forces that officials in Alxa try to create to turn ex-herders into *rou congrong* farmers is not the same as the "environment" that forestry engineers center. And yet their overlap patterns an "entangled excess" (Blaser and de la Cadena 2017, 186). In allowing "both diversity and cooperation," they are a boundary object that limns worlds, concerns, and motivations because of, and not despite, the disjunctures of their heterogeneous pairing (Star and Griesemer 1989, 393).

Holoparastic worldings and environmental governance emerge in the coordination of differences. They are a sympoeisis, a making-with, that begins from the proposition that nothing creates itself alone, that no plant relation or political system takes shape in isolation. Sympoeisis disrupts unidirectional vectors of cause and effect, subject and object. As Donna Haraway writes, it demands attention to worlds made in "intra-active relations in dynamic complex systems" (2016, 60). She encourages an ontology that refuses preexisting and bounded entities, in favor of non-innocent and non-optional relations that operate at once in multiple ontological registers (Chao 2018).

9. David Harvey, for instance, historicizes Deng Xiaoping and Reform and Opening in a global history of neoliberalism (2011). I build here on Lisa Rofel's aversion to the normative question of whether China has "arrived at the 'real' version of neoliberalism yet?" This is a "recurrent neocolonial [question] that [implies] infinite deferral" (2007, 7).

10. I emphasize the conceptual and political novelty of overgrazing for two reasons. First, the history of desertification as an environmental concept is closely linked with colonial land and population management especially in the Africa of the nineteenth century. Diana Davis shows how French colonial administrators in the Maghreb saw native "Arab nomads and their rapacious herds" (2004, 363) as agents of environmental destruction, creating "desert wastes" out of allegedly fertile lands. Armed with new concepts like carrying capacity and well-established tropes of oriental decadence, "overgrazing" operated as a colonial theory of coupled social and environmental degradation, and an alibi for various schemes of civilizational and landscape "improvement."

11. The biopolitical format of "make live, let die" (Foucault 2003, 241) and its murderous necropolitical revision, which links the achievement of sovereignty

with "exercising the right to kill" (Mbembe 2003, 12), are both evoked here. However, as we will see, the environmental experiments in this book pose power outside of the binary structure of living and dying, and the vitalist ontologies on which this account of the political is inevitably founded (see Cheah 2003).

12. According to the 2018 Annual Cashmere Market Report, published by The Schneider Group, Chinese producers account for 60 percent of global cashmere production. It notes that spikes in cashmere prices in that year were caused by "very strict environmental protection rules that led to a decreasing number of breeding goats" (Spina 2019).

13. Han families make up a significant proportion of (ex-)herding families in Alxa Banner. The vast majority of these families were originally farmers who migrated from neighboring Gansu province during the Great Famine of the late 1950s and learned to raise animals from Mongolian pastoralists. Ironically, many of them describe this migration as fleeing from desertification in Gansu.

14. Often preceded by disaster and accompanied by draconian neoliberal reforms, shock, argues Naomi Klein, has often been used in crisis as a cover for "extreme capitalist makeover" (2007, 7).

15. To modulate, following Gilbert Simondon, is different than to control. To modulate "is to mold in a continuously and perpetually variable way" (2020, 60). Highlighting modulation as a key feature of governance means foregoing the search for a theory of governance founded on the identification of a final, animating logic that would remain unchanged in its operation. This means that, in a sense, governance has no singular subject, but can be instead traced through the complex and unstable systems of elements that confront it. This is a possible meaning of "experiment" in this text.

16. See Judith Farquhar and Lili Lai's work (2014) on the salvage and sorting of minority ethnomedicines in Southwest China for another example of this.

17. This is an echo of the planned economy, except here, the local government is not setting quotas for supply as much as it is manipulating the conditions of demand (see Verdery 1996, chapter 1).

18. This sense of the real-time adjustment of conditions and parameters has important resonances with algorithmic and machine learning systems, which process new data in real time and continually retrain, recreating the conditions of their own data "environment." Thank you to Aakash Solanki for helping me make this connection.

19. Anthropologists of development have roundly critiqued the teleological underpinnings of hegemonic development paradigms, as well as the "antipolitics machine" of development that reframes political concerns into technical ones (Ferguson 1994). Here, I am more interested in how development shifts from a self-referencing logic and program into part of an experimental assemblage for making roots grow. That is, the process-teleology of development has been recircuited into an ecological tactic.

20. Anthropologists have written importantly of state and other attempts to cultivate entrepreneurial risk taking in populations figured as backward and intransigent in postsocialist transition (Dunn 2004), as paramount for neoliberal restructuring.

21. This is not terribly different from discussions that ponder the philosophical and technological conditions for the "intelligence" of artificial intelligences. It also edges close to orientalist tropes that have continually denied the possibility of Asian interior life and subjectivity, instead employing tropes of machines to explain Asian culture and personality (Rivera 2014, Rhee 2015). For a take that manages to hold both of these questions together, see the "Chinese Room Argument" (Searle 1980) about technological consciousness.

APPARATUS D. FIVE THOUSAND YEARS

1. An early version of this apparatus was introduced in the coda to an article in *Cultural Anthropology* (Zee 2017).

2. It is also, clearly, a shorthand to invoke the specificity of China for the Chinese American ethnographer, and therefore is always implicitly comparative in their conversations with me.

3. Conversely, her accounting of landscape change in social-ecological succession leaves little room for an accounting of actual changes in use patterns, especially in the last century. Scientists at her field station are especially worried over the depletion and contamination of the groundwater table, which would, indeed, throw the smooth cycle into more erratic loops.

CHAPTER 3. HOLDING PATTERNS

1. Parts of this article also appeared in an article of the same title in *Cultural Anthropology* (Zee 2017).

2. China's semiarid interior has for decades suffered massive desertification and has recently drawn scientific and journalistic attention for its epidemic of lake disappearances (Huo 2011).

3. On metaphorical extension, see Wagner 1986.

4. Environmental anthropologies also engage in and reflect on the temporal poetics of more-than-human assemblages. Whereas earlier works in ecological anthropology focused on the self-regulation of human-environmental systems (Rappaport 1984), contemporary environmental anthropology organizes modes of anticipation that complicate notions of timeless nature, from the idioms of endangerment and extinction that condition anticipatory nostalgia (Choy 2011; West 2006, 1–4) to the various modes of natural history that firmly

situate nature in time (Raffles 2002, White 1995). Together, these varied works signal an orientation toward specific modes of experience in time—historical, political, "natural"—emerging through potent nexuses, not all of which can be attributed to the work of human agents.

5. For instance, in what Nancy Oreskes and Erik Conway (2014), in their science fiction history of the future, call the Great Collapse, the inevitable climatic-political resolution of today's Great Acceleration, climate projections are also narrative forms. This narrative inversion, where environmental change rather than political action drives history, corresponds to their story of political inaction, with a single exception in a China that survives the collapse by virtue of illiberal, authoritarian rule as a conflicted agent of political proactivity.

6. Sand is a way of thinking through "short-circuits, unimaginable earlier, between the rhythm of history and that of *geohistory*" (Latour 2017, 45, italics in original).

7. "The invention of new forms of geo-social knowledge" (Tironi 2019) is a prospective enterprise of fabulating a vocabulary to grapple with stunning coming change. Here, I highlight how framing such metahistories—in genre, through reference to the progressions of history or the dynamics of an environmental process—describes a set of practices and forms through which time takes shape. "In what forms does the future manifest itself in the present" (Luhmann 1998, 64)?

8. Questions of environmental form have been important in asking, in anthropology, for ways of thinking about relations in time and space that are neither fully narrative nor linear in nature. Form, writes Kohn, manifests in "self-reinforcing pattern[s]" (2013, 180) that propagate across domains. Sand substantiates various temporal forms that powerfully interact with given political temporalities. In this propagation, politics' temporalities are "mediated and mutated by a form that is not exactly reducible to human events or landscapes" (2013, 183), even as they shape them.

9. Yi-Fu Tuan writes of the desert as a landform that breaks time's arrow through its negation of the generational temporalities of reproductive and sexual life (see also Edelman 2004 on "reproductive futurity"). Tuan figures the desert as "mineral purity free of death's remainders" and as a landscape that is synonymous with ruins. The desert, he writes only communicates "emptiness, stillness, indifference, divine imperturbability" (2001).

10. Reflections on statist time in recent anthropology trace specific modes of temporal experience as the ongoing conditions and achievements of political practices, running the gamut from "etatized" temporalities of waiting and boredom (Verdery 1996, 40; O'Neill 2017) to the repetitive cycles of failure and deferral in technocratic programs of governance (Ferguson 1994).

11. Li and Shapiro note that the elevation of "ecological civilization" (*shengtai wenming*) to an explicit goal of the Chinese Communist Party relies on a direct narrative throughline that cites and extends the official history of the

party. "Ecological civilization is described first and foremost as a continuation of China's developmental path under the leadership of the Communist Party . . . [and thus] a faithful reincarnation of Marx's theory of the stages of development with Chinese characteristics" (2020, 6).

12. Susan Buck-Morss has argued that socialisms of all kinds conjure their stakes in temporal terms, closely aligning political sovereignty with the coming to fruition of "dreamworlds of history" (2000), which share, with liberal political imaginaries, a presumption of progressivist development. Discussing China, Emily Ng, in her account of local rituals and cosmologies in Henan, suggests that time-making practices should be understood as "profound practices that incorporate and transform statist time" (2020, 69) rather than as simply oppositional or alterior to them.

13. Yanan Wang (2008) characterizes the county this way: "Minqin is surrounded by the Tengger and Badain Jaran Deserts on its east, west, and northern sides. At present, two large deserts are swallowing Minqin Oasis at a rate of about 10 meters per year."

14. *Jing*, or framed, conventionalized vista or scene, is a principle of Chinese garden and landscape architecture. It is "not simply a view of the landscape, but rather an active relationship between the changing environment and the human subject" (Jin 1998, 343; see also Zou 2011).

15. Anna Tsing, Andrew Mathews, and Nils Bubandt approach the Anthropocene through patchiness: "the uneven conditions of more-than-human livability in landscapes increasingly dominated by industrial forms" (2019, S186). They are after an ecological and existential attunement toward the uneven distribution of ecological emergences as a way of keying ethnographic method to Anthropocene emergences. Here, patchiness appears instead as a surgical spot-fix intervention for desertification engineering.

16. Johannes Fabian, in his discussion of anthropological figurations of Time, writes of succession in structuralist analysis, noting that diachrony is not history as such, but "the mere succession of semiological systems one upon another. Succession, strictly speaking, presupposes Time only in the sense of an extraneous condition affecting neither their synchronic nor their diachronic constitution" (2014, 56). That is, in both these modernist ecological paradigms and in mid-century structuralism, succession allows for an account of change as systemic evolution without reference to any actual, or in Fabian's terms, "physical" notion of Time *per se*.

17. Levenson argues that much of modern Chinese state political and philosophical thought has responded to the problem of how "to be modern and Chinese, that combination so desperately sought through a century of reformist and revolutionary exasperation as a seemingly immobile China and an all-too-kinetic West" (1968, 78). Levenson argues that one efficacy of the historical materialism of Marxist thinking was that it offered a way of retaining Chinese "tradition" as a

stage already being surpassed, and thus allowing for both an account of Chinese specificity and its transformation. Stagist and progressivist conceptions of history provided and continue to provide an official historiographic form, not only for making sense of Chinese pasts but also for angling toward futures framed as a succession of state-announced stages.

CHAPTER 4. PARTICULATE EXPOSURES

1. Orange is the second most severe official level of air pollution alert, after red. The announcement of red alerts is extremely rare, and the first was only announced in December 2015 (BBC News 2015). See Joseph Masco's discussion of color-coded threat levels in the United States (2014). He describes terror alerts, which are opaque in their actual calculation and have never, since they were introduced following 9/11, been set to low, as a technology for tuning public affect.

2. (US Embassy 2011). https://china.usembassy-china.org.cn/embassy-consulates/beijing/air-quality-monitor/extremely-high-levels-pm2-5-steps-reduce-exposure/

3. Thomas R. Johnson and Kathinka Fürst (2020) have cataloged some of the lively photography, painting, and performance art emerging in China in response to air pollution. Like Xi's walk, which captures haze as a question of meteorological and bodily contact through a sartorial decision, many of these pieces deploy clothing and accessories as atmospheric commentary, especially 3M surgical and painter's masks stitched into coverings, wedding dresses, and other items.

4. In retrospect, from the inside of the COVID-19 pandemic, that *un*-masking is figured as an act of forced solidarity is startling, especially as, in the United States, Trump's breathing-without-a-mask has been cynically politicized as an act of poorly conceived freedom. Here, liberal and socialist conceptions of the social collide in the possible political valences of mask wearing.

5. See Margaret Hillenbrand's (2020) discussion of public secrecy in China.

6. The COVID-19 pandemic has forefronted aerosols as a planetary mode of relation, as well as the condition of a micro-atmodynamic attention to how bodies are arranged, especially in indoor spaces. In each of these cases, it is not only the virus per se, but its aerosol properties and relations that have reopened questions of interpersonal, spatial, and international relations as matters of meteorological and atmospheric proximity.

7. Here, I invoke Stephen Collier's notion of a post-Soviet social (2011), which requires attentions to the proliferation of modes of social cohesion, ontology, and reorganization. In this, post-Soviet life in the former USSR could be posed not as a specific, epochal form, but through an analytic demand to attend to the proliferation of social possibilities and collectives taking shape. These post-Soviet

socials would both be understood as negotiations with the legacies of Soviet life and with the novel assemblages and relations of its aftermath.

8. Shigehisa Kuriyama reminds that "the history of the body is ultimately a history of ways of inhabiting the world" (2011, 237). Kuriyama argues that in classical Chinese understandings of the body, the body is not simply immersed in weather and vulnerable to its caprices. It is itself an atmospheric fold, a temporary topology of wind bent on itself until its inevitable release. Illness, in classical Chinese medical thought, would be the effect the atmospheric systems of the body and its environment falling out of phase, making maladjusted contact through openings—breath and pores. And in this sense, the relation between bodies and their mediums is a question of the contact between interacting atmospheric dynamics, the interface of the weather with itself across a temporary corporeal sheath.

9. Chen's original title for the piece is simply *"Wumai,"* or haze.

10. From a Weibo post dated November 1, 2011.

11. From a Weibo post dated November 3, 2011.

APPARATUS E. WILDFIRES

1. The Hong Kong–based decolonial left collective Lausan has warned against a utopian romance of these protests, noting that they are largely predicated on the reliance on invisibilized and precarious foreigner workforces, especially domestic workers from Southeast Asia who often work in extremely unstable conditions (T and Squatting 2020; see Ong (2006) chapter: "A Biocartography: Maids, Neoslavery, and NGOs"). These protests also express a range of political perspectives that run the gamut of leftist internationalism to neo-fascist appeals inviting military and other intervention from the US right wing (Zhao 2020).

2. June 4, 2020, was the first year, amid COVID-19 and a new Hong Kong national security law, that mass memorial ceremonies for the Tiananmen Square gathering were banned in Hong Kong.

3. Kristen Simmons's description of the police and military use of gas and water against indigenous bodies at Standing Rock and the Mni Sose, the Missouri River, emphasizes that "suspension is a condition of settler colonialism." Defilements of breathable air can be understood, following Simmons, as a tactic of "asphyxiation" through which "colonial governmentality necessarily strangulates other forms of relationality and coalition building" (2017). These scenes of relationality and collectivity as atmospheric entailments come to a head in the next chapter.

4. These tactics of tear gas choreography and containment themselves articulate a global geography of protest tactics, especially as gaseous agents have found widespread deployment in suppressing domestic protest, even as they are banned internationally as a medium of chemical warfare. These tactics appeared again in protests across the United States and beyond that erupted and continue against

white supremacy and anti-Black police brutality in the aftermath of the deaths of George Floyd, Breonna Taylor, and countless named and unnamed others.

5. In the years since the event, these hashtags have remained active. They repost photos from before lockdowns, and, troubling, they have revealed a spectrum of political positions, including some that express a proto-fascist and xenophobic nationalism.

6. For more, see Shan Huang's reflections on fieldwork in and of the protest movements. He argues that the demand for the "mutual care for strangers" in the protest movement has forged a collective identity of citizens as "members of a larger moral-emotional community" (2020).

CHAPTER 5. CITY OF CHAMBERS

1. The canned air suggests an immediate parallel with the opening scenes of Mel Brooks's 1987 classic science-fiction parody, *Spaceballs*, where the planet Spaceball has wasted all of its air. In the scene, President Skroob, in his sleek office, dismisses inquiries from journalists on the phone calling about "rumors" that the planet is suffering from an air shortage. "Shithead," he says, hanging up the phone and reaching into a desk drawer filled with cans of "Perri-air," naturally sparkling, salt-free, and canned on the neighboring planet Druidia. He opens and inhales eagerly and desperately before being interrupted by a video-call from a military official. See the clip here: www.youtube. com/watch?v=SiabeNR_qoU.

2. Canetti lingers, a specter in much writing on breath and space. His atmospheric urbanism attends to bodies as gradient airspaces, a space of both mixture and its limits. He writes, "The big city is as full of such breathing-spaces as of individual people. Now none of these people is like the next, each is a kind of cul-de-sac; and just as their splintering makes up the chief attraction and chief distress of life, so too one could also lament the splintering of the atmosphere" (1979, 13). The lament over splintering, however, threatens to lead too easily into another cul-de-sac if our attention ends with the critique of privatization, folding the atmosphere into another site in the agency of capital, or into the site of a collapse of a normative democracy. Dipesh Chakrabarty, in a closely related argument, argues that "the analytics of capital (or of the market), while necessary, are insufficient instruments in helping us come to grips with anthropogenic climate change" (2014, 4).

3. In any case, in what sense is the air an analog for the public if one refuses the constitution of air primarily as a resource rather than a medium?

4. The fantasy of hermetic sealing, notes Gokce Gunel, analogizes life on Earth to a spacefaring vessel. The spaceship signifies "enclosure, archiving, selection, hierarchy, movement, and—most importantly—the maintenance of strict boundaries between interior and exterior spaces" (2019, 41).

5. See Calvillo's site for real-time updates to this project, at http://intheair.es /index.html.

6. Tim Choy describes the mixture of dismayed identification and revulsion that some Chinese American anthropologists feel when they encounter their "imagined doppelgangers, the well-paid expatriates (including those of Chinese descent)" (2011, 142).

7. I owe this lovely insight to Seth Denizen.

8. A major company operating in this field, and which supplied the International School of Beijing is ASATI: Air Structures American Technologies, Inc., based in New York. At the time of writing, the company's website (https://www .asati.com/) has refocused around providing air architecture and enclosures for pop-up COVID-19 quarantine facilities. The identification of air as a rapid architectural medium, for deployment domestically in the United States by the Federal Emergency Management Administration (FEMA) conjures other histories of disaster shelter and atmospheric attunement. See Shapiro 2015.

9. I thank Nicholas D'Avella for turning me on to this quote. More on the architectural logic and form of different building materials is in his book *Concrete Dreams: Practice, Value, and Built Environments in Argentina* (2019).

10. In Mao's speeches and writings on the fable, the Foolish Old Man toils to remove mountains, heavy "like a dead weight on the Chinese people. One is imperialism and one is feudalism" (1945).

APPARATUS F. A SINOCENE

1. All translations of Pan Yue are mine. "On Socialist Ecological Civilization" was widely disseminated, and can be read as an attempt to lend a retroactive theoretical and ideological coherence to the notion of "ecological civilization," which, for some time has been circulating as a semi-official slogan. Ecological civilization should be understood itself as part of a conventionalized list of other invocations of "civilization" (*wenming*), which scholars have analyzed in the context of official anxieties over the problem of low human "quality" (*suzhi*) in China since the 1990s as "neoliberalism reified" (Kipnis 2007; see also Anagnost 2004, Tomba 2009).

2. The historical ontologization of capitalism as a necessary engine of ecological destruction has been an important feature in multiple political ecologies and critiques of capitalism (Moore 2015).

CHAPTER 6. DOWNWINDS

1. I am deeply indebted to conversations with Dr. Ian Faloona at UC Davis that helped me learn about the worlds of transpacific aerosol research on the

North American West Coast. I first learned about his work in journalistic cover-age of his lab's findings (Ortiz 2015). Much of this work considers aerosol trans-port in relation to ozone transport. Some material from this chapter appears in the edited volume *Voluminous States: Sovereignty, Materiality, and the Territo-rial Imagination* (Zee 2020b).

2. The Mt. Tamalpais site is one of several aerosol research stations with RDI's on the Pacific Coast. Others are distributed through federal lands and national parks through the IMPROVE Project.

3. "Rotating drum impactors (RDI) are cascade-type impactors used for size- and time-resolved aerosol sampling, mostly followed by spectrometric analysis of the deposited material. They are characterized by one rectangular nozzle per stage and are equipped with an automated stepping mechanism for the impac-tion wheels" (Bukowiecki et al. 2009, 891)

4. Baselines, as a point of originary reference that is not identified with the absolute, are a way of denoting norms as well as gesturing at the flexibility in their determination. They appear in wide ranges of quantitative knowledge making, especially in economics, but also in disciplines as diverse as toxicology and studies of human behavior.

5. The full text of this section of the Clean Air Act is available here: https://www.govinfo.gov/content/pkg/USCODE-2013-title42/html/USCODE-2013-title42-chap85-subchapI-partD-subpart1-sec7509a.htm.

6. Consider this alongside Mel Y. Chen's discussion of the inherent risks of Chinese imports to U.S. consumers. Chen traces a durable narrative that pitches "these environmental threats neither as 'acts of God' nor as products of global industrialization, but as invasive dangers into the U.S. territory from other national territories. These environmental toxins were supposed to be 'there,' but were found 'here'" (2011, 267), in the United States. Chen's account is one of mat-ter out of place. The notion of a baseline also refuses the naturalism of an "act of God," but reconceives the naturalism of national borders for a post-natural mea-sure of atmospheric normalcy.

7. Incidentally, the dust event in question is the same April 2001 event described in this book's introduction. See NASA's description of the event at https://science.nasa.gov/science-news/science-at-nasa/2001/ast17may_1. "North America has been sprinkled with a dash of Asia! A dust cloud from China crossed the Pacific Ocean recently and rained Asian dust from Alaska to Florida."

8. Recent attempts at specifying the terms of intra-Asian relations contend with the fraught genealogy of Asia as a cartographic and civilizational fiction. Scholar of the Chinese New Left Wang Hui has argued that Asia, is first and foremost a European invention, coming out of eighteenth-century European philosophical accounts of universal history. In search of the coordinate of an Asia that arises from Asia, Wang derides "Asian" Asias derivative of European political and philosophical thought, especially Imperial Japan's definition of a

Greater East Asia Co-prosperity Sphere as the natural regional growing space for the expansion of a Japan-centric Asia (2011). Gayatri Spivak has posed the question of Asia against the geopolitical demands of US foreign policy and the Cold War area studies paradigm, and without reprising the political project of various Pan-Asianisms, has focused instead on Asian plurality (2008).

9. Imports trigger an anxiety of being downwind, brewed in the braiding of inbound weather patterns and as versions of a composite Chinese assault on an already waning sense of American sovereignty (Brown 2014) percolating as externalities of globalization. In this, dust appears as redundant, a mocking planetary iteration of a reorientation of global relations around a Sino-American axis, measured in outsourced production and registered affectively in the exhaustion, anxiety, and "excitements" of supply-chain capitalism (Tsing 2009, 149).

10. This is akin to what Eyal Weizman has described, in his writings on political volume and occupation in Israel-Palestine as a politics of verticality (2017).

11. See Vivian Choi's discussion of atmospheric attunement and emergent vernacular regimes of sensibility for anticipating risk in the aftermath of the 2005 Indian Ocean tsunami (2015).

12. There are at least four groups in Seoul with tree-planting projects in Inner Mongolia, which is popularly understood in Korean media as the principal source of yellow dusts. All of these are not-for-profit groups, with formal and informal ties with the Korean Forestry Administration. One is part of a foundation and corporate social responsibility outfit for an international paper conglomerate.

13. See Tim Choy's description of Hong Kong air quality and the upwind relocation of polluting factories from Hong Kong to China's Guangdong province (2011).

14. This specific question of international air pollution as a function of Chinese production for export renders the "race to the bottom" of factory relocation and labor exploitation in distinctly geo-atmospheric terms (Lin et al. 2014).

15. Because this declaration of friendship refuses the designation of the enemy as the definition of the political as such (Schmitt 2007), it takes on a distinctly meteorological character.

APPARATUS G. MONSTERS

1. An earlier description of iron and mercury in dust appears in *Feral Atlas* (Zee 2020c).

2. Ocean iron seeding has been proposed as a geo engineering technique for increasing the carbon sequestration capacity of the ocean by inducing plankton bloom. Several studies have appraised the benefits of whale excreta for ocean carbon storage (see Pershing 2010, Lavery 2010).

3. For a contemporary study of baleen whales and the iron cycle in the southern Pacific Ocean, sea Nicol et al. 2010.

4. See also Alex Blanchette's discussion of industrial meat production and the chemical-excrement clouds that rise from lagoons of swine and steer waste. He writes, "Dust and other forms of fugitive waste become legible as a curious kind of pollutant that does not merely settle onto the surfaces of biological bodies so much as it merges with and passes through them" (2019, 82).

5. Stefanie Graeter gets the last word in this book. In her ethnography of lead contamination and logistics in Peru, Graeter attends to "assemblages of bodies, heavy metals, and value mediated by extractive infrastructures" (2020, 22). In its imbrication into bodies, neurologies, and ecologies, the lead logistics chain is not only a vector of damage but also a concrete (and crumbling) infrastructure whose operation pre-echoes the emergence of new modes of human and more-than-human embodiment. Like mountain lions, residents along lead transport corridors "continuously accumulate [heavy metals] in their brains, bones, and tissue" becoming a living "infrastructure of toxic storage for contemporary extractive capitalism" (2020, 24).

References

Abe, Kōbō. 1991. *The Woman in the Dunes.* New York: Vintage Books.

Acta Scientiarum Naturalium Universitatis Intramongolicae. 1960. "Cistanche Deserticola." *Acta Scientiarum Naturalium Universitatis Intramongolicae* (1): 63.

Adams, Vincanne, Michelle Murphy, and Adele E. Clarke. 2009. "Anticipation: Technoscience, Life, Affect, Temporality." *Subjectivity* 28 (1): 246–65. https://doi.org/10.1057/sub.2009.18.

Adey, Peter. 2014. *Air.* London, UK: Reaktion Books.

Agard-Jones, Vanessa. 2012. "What the Sands Remember." *GLQ: A Journal of Lesbian and Gay Studies* 18 (2–3): 325–46. https://doi.org/10.1215/10642684 -1472917.

Agrawal, Arun. 2005. "Environmentality: Community, Intimate Government, and the Making of Environmental Subjects in Kumaon, India." *Current Anthropology* 46 (2): 161–90.

Ahlers, Anna, and Mette Halskov Hansen. 2019. "Air Pollution: How Will China Win Its Self-Declared War Against It?" In *Routledge Handbook of Environmental Policy in China,* edited by Eva Sternfeld. New York: Routledge.

Ahuja, Neel. 2016. *Bioinsecurities: Disease Interventions, Empire, and the Government of Species.* Durham, NC: Duke University Press.

Alaimo, Stacy. 2016. *Exposed: Environmental Politics and Pleasures in Post-human Times.* Minneapolis: University of Minnesota Press.

Amrith, Sunil S. 2018. *Unruly Waters: How Rains, Rivers, Coasts, and Seas Have Shaped Asia's History*. New York: Basic Books.

Anagnost, Ann. 1997. *National Past-Times: Narrative, Representation, and Power in Modern China*. Durham, NC: Duke University Press.

——. 2004. "The Corporal Politics of Quality (Suzhi)." *Public Culture* 16 (2): 189–208.

Anand, Nikhil. 2017. *Hydraulic City: Water and the Infrastructures of Citizenship in Mumbai*. Durham, NC: Duke University Press.

Anand, Nikhil, Akhil Gupta, and Hannah Appel, eds. 2018. *The Promise of Infrastructure*. Durham, NC: Duke University Press.

Andrews, Steven Q. 2008. "Inconsistencies in Air Quality Metrics: 'Blue Sky' Days and PM10 Concentrations in Beijing." *Environmental Research Letters* 3 (3): 034009. https://doi.org/10.1088/1748-9326/3/3/034009.

Bagnold, Ralph A. (1941) 2005. *The Physics of Blown Sand and Desert Dunes*. Mineola, NY: Dover Publications.

——. 2010. *Libyan Sands: Travel in a Dead World*. London: Eland.

Ballestero, Andrea. 2019a. *A Future History of Water*. Durham, NC: Duke University Press.

——. 2019b. "Aquifers (or, Hydrolithic Elemental Choreographies)." *Cultural Anthropology Online* (blog). June 27, 2019. https://culanth.org/fieldsights/aquifers-or-hydrolithic-elemental-choreographies.

——. 2019c. "Touching with Light; or, How Texture Recasts the Sensing of Underground Water." *Science, Technology, & Human Values* 44 (5): 762–85. https://doi.org/10.1177/0162243919858717.

Barad, Karen. 2007. *Meeting the Universe Halfway: Quantum Physics and the Entanglement of Matter and Meaning*. Durham, NC: Duke University Press.

Barry, J. L. 2001. "The Pacific Dust Express." NASA. https://science.nasa.gov/science-news/science-at-nasa/2001/ast17may_1

Baskin, Carol C., and Jerry M. Baskin. 2014. *Seeds: Ecology, Biogeography, and Evolution of Dormancy and Germination*. 2nd edition. San Diego, CA: Elsevier/AP.

Battaglia, Debbora, David Valentine, and Valerie Olson. 2015. "Relational Space: An Earthly Installation." *Cultural Anthropology* 30 (2): 245–56. https://doi.org/10.14506/ca30.2.07.

BBC. 2015. "China Pollution: First Ever Red Alert in Effect in Beijing," December 8, 2015. https://www.bbc.com/news/world-asia-china-35026363.

Benjamin, Walter. 2006. "On the Concept of History." In *Selected Writings, 1938–1940*. Edited by Howard Eiland and Michael W. Jennings. Cambridge, MA: Harvard University Press.

Bennett, Jane. 2010. *Vibrant Matter: A Political Ecology of Things*. Durham, NC: Duke University Press.

——. 2020. *Influx and Efflux: Writing Up with Walt Whitman*. Durham, NC: Duke University Press. https://doi.org/10.2307/j.ctv11smzcr.

Berlant, Lauren. 2016. "The Commons: Infrastructures for Troubling Times." *Environment and Planning D: Society and Space* 34 (3): 393–419. https://doi .org/10.1177/0263775816645989.

Berman, Marshall. 1988. *All That Is Solid Melts into Air: The Experience of Modernity.* New York: Viking Penguin.

Bernstein, Anya. 2019. *The Future of Immortality: Remaking Life and Death in Contemporary Russia.* Princeton, NJ: Princeton University Press.

Biehl, João, and Peter Locke, eds. 2017. "Unfinished." In *Unfinished: The Anthropology of Becoming,* ix–xiii. Durham, NC: Duke University Press.

Biggs, David A. 2018. *Footprints of War: Militarized Landscapes in Vietnam.* Seattle: University of Washington Press.

Billé, Franck. 2016. *Sinophobia: Anxiety, Violence, and the Making of Mongolian Identity.* Honolulu, HI: University of Hawai'i Press.

———, ed. 2020. *Voluminous States: Sovereignty, Materiality, and the Territorial Imagination.* Durham, NC: Duke University Press.

Blaikie, Piers Macleod, and Harold Chillingworth Brookfield. 2015. *Land Degradation and Society.* New York: Routledge.

Blake, William. 1950. "Auguries of Innocence." In *Poets of the English Language.* New York: Viking Press. https://www.poetryfoundation.org/poems/43650 /auguries-of-innocence.

Blaser, Mario, and Marisol de la Cadena. 2017. "The Uncommons: An Introduction." *Anthropologica* 59 (2): 185–93. https://doi.org/10.3138/anth.59.2.t01.

Borges, José Luis. 1981. "On Exactitude in Science." In *A Universal History of Infamy,* translated by Norman T. Giovanni. New York: Penguin.

Boyer, Dominic. 2019. *Energopolitics: Wind and Power in the Anthropocene.* Durham, NC: Duke University Press.

Bray, David. 2005. *Social Space and Governance in Urban China: The Danwei System from Origins to Reform.* Stanford, CA: Stanford University Press.

Brooks, Mel. 1987. *Spaceballs.* Metro-Goldwyn-Mayer.

Brown, Wendy. 2014. *Walled States, Waning Sovereignty.* New York: Zone Books.

Buck-Morss, Susan. 2002. *Dreamworld and Catastrophe: The Passing of Mass Utopia in East and West.* Cambridge, MA: MIT Press.

Bukowiecki, Nicolas, Agnes Richard, Markus Furger, Ernest Weingartner, Myriam Aguirre, Thomas Huthwelker, Peter Lienemann, Robert Gehrig, and Urs Baltensperger. 2009. "Deposition Uniformity and Particle Size Distribution of Ambient Aerosol Collected with a Rotating Drum Impactor." *Aerosol Science and Technology* 43 (9): 891–901. https://doi.org/10.1080 /02786820903002431.

Bulag, Uradyn E. 2002. *The Mongols at China's Edge: History and the Politics of National Unity.* Lanham, MD: Rowman & Littlefield Publishers.

Bunzl, Matti. 1993. *From Historicism to Historical Particularism: Franz Boas and the Tradition of 19th Century German Anthropology and Linguistics.* M. Bunzl.

Butler, Judith. 2004. *Undoing Gender.* New York & London: Routledge.
———. 2015. *Notes Toward a Performative Theory of Assembly.* Cambridge, MA: Harvard University Press.
Byrnes, Corey J. 2018. *Fixing Landscape: A Techno-Poetic History of China's Three Gorges.* New York: Columbia University Press.
———. 2020. "Transpacific Maladies." *Social Text* 38 (3): 1–26.
Callison, Candis. 2014. *How Climate Change Comes to Matter: The Communal Life of Facts.* Durham, NC: Duke University Press.
Canetti, Elias. 1979. *The Conscience of Words.* New York: Seabury Press.
Caple, Zachary. 2017. "Holocene in Fragments: A Critical Landscape Ecology of Phosphorus in Florida." PhD dissertation, University of California, Santa Cruz.
Carse, Ashley. 2014. *Beyond the Big Ditch: Politics, Ecology, and Infrastructure at the Panama Canal.* Cambridge, MA: MIT Press.
Carse, Ashley, Jason Cons, and Townsend Middleton. 2018. "Chokepoints." *Limn* 10. https://limn.it/articles/preface-chokepoints/.
Chakrabarty, Dipesh. 2009. "The Climate of History: Four Theses." *Critical Inquiry* 35 (2): 197–222. https://doi.org/10.1086/596640.
———. 2014. "Climate and Capital: On Conjoined Histories." *Critical Inquiry* 41 (1): 1–23. https://doi.org/10.1086/678154.
Chan, C. Y., X. D. Xu, Y. S. Li, K. H. Wong, G. A. Ding, L. Y. Chan, and X. H. Cheng. 2005. "Characteristics of Vertical Profiles and Sources of PM2.5, PM10 and Carbonaceous Species in Beijing." *Atmospheric Environment* 39 (28): 5113–24. https://doi.org/10.1016/j.atmosenv.2005.05.009.
Chao, Sophie. 2018. "In the Shadow of the Palm: Dispersed Ontologies among Marind, West Papua." *Cultural Anthropology* 33 (4): 621–49. https://doi.org/10.14506/ca33.4.08.
Cheah, Pheng. 2003. *Spectral Nationality: Passages of Freedom from Kant to Postcolonial Literatures of Liberation.* New York: Columbia University Press.
Chen, Fahu, Shengqian Chen, Xu Zhang, Jianhui Chen, Xin Wang, Evan J. Gowan, Mingrui Qiang, et al. 2020. "Asian Dust-Storm Activity Dominated by Chinese Dynasty Changes since 2000 BP." *Nature Communications* 11 (1): 992. https://doi.org/10.1038/s41467-020-14765-4.
Chen, Kuan-Hsing. 2010. *Asia as Method: Toward Deimperialization.* Durham, NC: Duke University Press.
Chen, M. Y. 2011. "Toxic Animacies, Inanimate Affections." *GLQ: A Journal of Lesbian and Gay Studies* 17 (2–3): 265–86. https://doi.org/10.1215/10642684-1163400.
Chen, Nancy N. 2003. *Breathing Spaces: Qigong, Psychiatry, and Healing in China.* New York: Columbia University Press.
Chen, Qiufan. 2015. "The Smog Society." Translated by Ken Liu and Carmen Yiling Yan. *Lightspeed Magazine,* 2015. http://www.lightspeedmagazine.com/fiction/the-smog-society/.

Chen, Siyu, Jianping Huang, Yun Qian, Chun Zhao, Litai Kang, Ben Yang, Yong Wang, et al. 2017. "An Overview of Mineral Dust Modeling over East Asia." *Journal of Meteorological Research* 31 (4): 633–53. https://doi.org/10.1007/s13351-017-6142-2.

Chen, Y., A. Ebenstein, M. Greenstone, and H. Li. 2013. "Evidence on the Impact of Sustained Exposure to Air Pollution on Life Expectancy from China's Huai River Policy." *Proceedings of the National Academy of Sciences* 110 (32): 12936–41. https://doi.org/10.1073/pnas.1300018110.

Cho, Mun Young. 2013. *The Specter of "The People": Urban Poverty in Northeast China*. Ithaca, NY: Cornell University Press.

Choi, Vivian Y. 2015. "Anticipatory States: Tsunami, War, and Insecurity in Sri Lanka." *Cultural Anthropology* 30 (2): 286–309. https://doi.org/10.14506/ca30.2.09.

Chow, Rey. 2012. *Entanglements: Or Transmedial Thinking about Capture*. Durham, NC: Duke University Press.

Choy, Timothy K. 2011. *Ecologies of Comparison: An Ethnography of Endangerment in Hong Kong*. Durham, NC: Duke University Press.

———. 2016. "Distribution." Theorizing the Contemporary. *Fieldsights*, January 21. https://culanth.org/fieldsights/distribution.

———. 2018. "Tending to Suspension: Abstraction and Apparatuses of Atmospheric Attunement in Matsutake Worlds." *Social Analysis* 62 (4): 54–77. https://doi.org/10.3167/sa.2018.620404.

Choy, Timothy K., and Jerry C. Zee. 2015. "Condition—Suspension." *Cultural Anthropology* 30 (2): 210–23. https://doi.org/10.14506/ca30.2.04.

Chu, Julie Y. 2010. *Cosmologies of Credit: Transnational Mobility and the Politics of Destination in China*. Durham, NC: Duke University Press.

———. 2014. "When Infrastructures Attack: The Workings of Disrepair in China." *American Ethnologist* 41 (2): 351–67. https://doi.org/10.1111/amet.12080.

Chuang, P. Y., R. M. Duvall, M. M. Shafer, and J. J. Schauer. 2005. "The Origin of Water Soluble Particulate Iron in the Asian Atmospheric Outflow." *Geophysical Research Letters* 32 (7): n/a-n/a. https://doi.org/10.1029/2004GL021946.

Clarke, Bruce. 2020. *Gaian Systems: Lynn Margulis, Neocybernetics, and the End of the Anthropocene*. Minneapolis: University of Minnesota Press.

Clements, Frederic E. 1916. *Plant Succession: An Analysis of the Development of Vegetation*. Washington, DC: Carnegie Institute of Washington.

Coen, Ross Allen. 2014. *Fu-Go: The Curious History of Japan's Balloon Bomb Attack on America*. Lincoln: University of Nebraska Press.

Cohen, Jeffrey Jerome, ed. 2013. *Prismatic Ecology: Ecotheory beyond Green*. Minneapolis: University of Minnesota Press.

Collier, Stephen J. 2009. "Topologies of Power: Foucault's Analysis of Political Government beyond 'Governmentality.'" *Theory, Culture & Society* 26 (6): 78–108. https://doi.org/10.1177/0263276409347694.

———. 2011. *Post-Soviet Social: Neoliberalism, Social Modernity, Biopolitics.* Princeton, NJ: Princeton University Press.

Confucius. 2003. *Confucius's Analects: With Selection from Traditional Commentaries.* Translated by Edward G. Slingerland. Indianapolis, IN: Hackett Pub. Co.

Cons, Jason, and Michael Eilenberg, eds. 2019. *Frontier Assemblages: The Emergent Politics of Resource Frontiers in Asia.* Antipode Book Series. Hoboken, NJ: John Wiley & Sons.

Cowles, Henry Chandler. 1899. "The Ecological Relations of the Vegetation on the Dunes of Lake Michigan." *Botanical Gazette* 27 (2): 95–117.

Cox, Aimee Meredith. 2015. *Shapeshifters: Black Girls and the Choreography of Citizenship.* Durham, NC: Duke University Press.

Cronon, William. 1996. "The Trouble with Wilderness: Or, Getting Back to the Wrong Nature." *Environmental History* 1 (1): 7. https://doi.org/10.2307/3985059.

Dai, Jinhua. 2018. "Introduction." In *After the Post-Cold War: The Future of Chinese History*, by Jinhua Dai, edited by Lisa Rofel, translated by Jie Li. Durham, NC: Duke University Press.

D'Avella, Nicholas. 2014. "Ecologies of Investment: Crisis Histories and Brick Futures in Argentina." *Cultural Anthropology* 29 (1): 173–99. https://doi.org/10.14506/ca29.1.10.

———. 2019. *Concrete Dreams: Practice, Value, and Built Environments in Post-Crisis Buenos Aires.* Durham, NC: Duke University Press.

Davis, Diana K. 2004. "Desert 'Wastes' of the Maghreb: Desertification Narratives in French Colonial Environmental History of North Africa." *Cultural Geographies* 11 (4): 359–87. https://doi.org/10.1191/1474474004eu313oa.

———. 2007. *Resurrecting the Granary of Rome: Environmental History and French Colonial Expansion in North Africa.* Athens: Ohio University Press.

———. 2016. *The Arid Lands: History, Power, Knowledge.* Cambridge, MA: The MIT Press.

———. 2017. "Desertification." In *International Encyclopedia of Geography: People, the Earth, Environment and Technology*, edited by Douglas Richardson, Noel Castree, Michael F. Goodchild, Audrey Kobayashi, Weidong Liu, and Richard A. Marston, 1–10. Oxford, UK: John Wiley & Sons. https://doi.org/10.1002/9781118786352.wbieg0560.

Day, Iyko. 2016. *Alien Capital: Asian Racialization and the Logic of Settler Colonial Capitalism.* Durham: Duke University Press.

De Beauvoir, Simone. 2001. *The Long March: An Account of Modern China.* London: Phoenix.

De Certeau, Michel. 2013. *The Practice of Everyday Life.* Berkeley: University of California Press.

De la Cadena, Marisol. 2015. *Earth Beings: Ecologies of Practice Across Andean Worlds.* Durham, NC: Duke University Press.

De Landa, Manuel. 1997. *A Thousand Years of Nonlinear History*. New York: Zone Books.

Deleuze, Gilles, and Felix Guattari. 2000. *A Thousand Plateaus: Capitalism and Schizophrenia*. London: Continuum.

Demos, T.J. 2019. "Climate Control: From Emergency to Emergence." *E-Flux* 104 (November). https://www.e-flux.com/journal/104/299286/climate -control-from-emergency-to-emergence/.

Derrida, Jacques. 1993. "Politics of Friendship." *American Imago* 50 (3): 353–91.

———. 2006. *Specters of Marx: The State of the Debt, the Work of Mourning and the New International*. Routledge Classics. New York: Routledge.

———. 2008. *The Animal That Therefore I Am*. Trans. Marie-Louise Mallet. New York: Fordham University Press.

Domagalski, Joseph, Michael S. Majewski, Charles N. Alpers, Chris S. Eckley, Collin A. Eagles-Smith, Liam Schenk, and Susan Wherry. 2016. "Comparison of Mercury Mass Loading in Streams to Atmospheric Deposition in Watersheds of Western North America: Evidence for Non-Atmospheric Mercury Sources." *Science of The Total Environment* 568 (October): 638–50. https:// doi.org/10.1016/j.scitotenv.2016.02.112.

Douglas, Mary. 2005. *Purity and Danger: An Analysis of Concepts of Pollution and Taboo*. London : Routledge.

Duca, Laura. 2016. "Donald Trump Is Gaslighting America." *Teen Vogue*, December 10, 2016. https://www.teenvogue.com/story/donald-trump-is -gaslighting-america.

Dunn, Elizabeth C. 2004. *Privatizing Poland: Baby Food, Big Business, and the Remaking of Labor*. Ithaca, NY: Cornell University Press.

The Economist. 2013. "Small Beginnings." *The Economist Microblogs* (blog). April 6, 2013. https://www.economist.com/special-report/2013/04/06/small -beginnings.

Economy, Elizabeth. 2005. *The River Runs Black: The Environmental Challenge to China's Future*. Ithaca, NY: Cornell University Press.

Edelman, Lee. 2004. *No Future: Queer Theory and the Death Drive*. Series Q. Durham, NC: Duke University Press.

Edwards, Paul N. 2002. "Infrastructure and Modernity: Force, Time, and Social Organization in the History of Sociotechnical Systems." In *Modernity and Technology*, edited by Thomas J. Misa, Philip Brey, and Andrew Feenberg. Cambridge, MA: MIT Press.

———. 2006. "Meteorology as Infrastructural Globalism." *Osiris* 21 (1): 229–50. https://doi.org/10.1086/507143.

———. 2013. *A Vast Machine: Computer Models, Climate Data, and the Politics of Global Warming*. Cambridge, MA: MIT Press.

Elvin, Mark. 1998. "Who Was Responsible for the Weather? Moral Meteorology in Late Imperial China." *Osiris* 12: 213–37.

———. 2006. *The Retreat of the Elephants: An Environmental History of China.* New Haven, CT: Yale University Press.

Embassy Beijing. 2009. "Embassy Air Quality Tweets Said to 'Confuse' Chinese Public. Wikileaks Cable: 09Beijing 1945_a." https://wikileaks.org/plusd /cables/09BEIJING1945_a.html.

Engelmann, Sasha. 2015. "Toward a Poetics of Air: Sequencing and Surfacing Breath." *Transactions of the Institute of British Geographers* 40 (3): 430–44. https://doi.org/10.1111/tran.12084.

Ewing, Stephanie A., John N. Christensen, Shaun T. Brown, Richard A. Vancuren, Steven S. Cliff, and Donald J. Depaolo. 2010. "Pb Isotopes as an Indicator of the Asian Contribution to Particulate Air Pollution in Urban California." *Environmental Science & Technology* 44 (23): 8911–16. https://doi.org/10.1021/es101450t.

Fabian, Johannes. 2014. *Time and the Other: How Anthropology Makes Its Object.* New York: Columbia University Press.

Faier, Lieba, and Lisa Rofel. 2014. "Ethnographies of Encounter." *Annual Review of Anthropology* 43: 363–77.

Fang, Yang. 2014. "Pollution Study to Use Smog Chambers." *Global Times*, February 3, 2014. www.globaltimes.cn/ content/845678.shtml.

Farquhar, Judith, and Lili Lai. 2014. "Information and Its Practical Other: Crafting Zhuang Nationality Medicine." *East Asian Science, Technology and Society* 8 (4): 417–37. https://doi.org/10.1215/18752160-2721450.

Farquhar, Judith, and Qicheng Zhang. 2012. *Ten Thousand Things: Nurturing Life in Contemporary Beijing.* New York: Zone Books.

Fedman, David. 2018. "The Ondol Problem and the Politics of Forest Conservation in Colonial Korea." *Journal of Korean Studies* 23 (1): 25–64. https://doi .org/10.1215/21581665-4339053.

———. 2020. *Seeds of Control: Japan's Empire of Forestry in Colonial Korea.* Seattle: University of Washington Press.

Fei, Xiaotong. 1992. *From the Soil, the Foundations of Chinese Society.* Translated by Gary G. Hamilton and Zheng Wang. Berkeley: University of California Press.

Feigenbaum, Anna. 2017. *Tear Gas: From the Battlefields of World War I to the Streets of Today.* London: Verso.

Feng, Xiao, Qi Li, Yajie Zhu, Jingjie Wang, Heming Liang, and Ruofeng Xu. 2014. "Formation and Dominant Factors of Haze Pollution over Beijing and Its Peripheral Areas in Winter." *Atmospheric Pollution Research* 5 (3): 528–38. https://doi.org/10.5094/APR.2014.062.

Ferguson, James. 1994. *The Anti-Politics Machine: "Development," Depoliticization, and Bureaucratic Power in Lesotho.* Minneapolis: University of Minnesota Press.

Field, Jason P, Jayne Belnap, David D Breshears, Jason C Neff, Gregory S Okin, Jeffrey J Whicker, Thomas H Painter, Sujith Ravi, Marith C Reheis, and

Richard L Reynolds. 2010. "The Ecology of Dust." *Frontiers in Ecology and the Environment* 8 (8): 423–30. https://doi.org/10.1890/090050.

Fisch, Michael. 2018. *An Anthropology of the Machine: Tokyo's Commuter Train Network*. Chicago: University of Chicago Press.

Fleming, James Rodger. 2012. *Fixing the Sky: The Checkered History of Weather and Climate Control*. New York: Columbia University Press.

Flikke, Rune. 2016. "Enwinding Social Theory: Wind and Weather in Zulu Zionist Sensorial Experiences." *Social Analysis* 60 (3). https://doi.org/10.3167/sa.2016.600306.

Flusser, Vilém. 2013. *Post-history*. Edited by Siegfried Zielinski. Translated by Rodrigo Maltez Novaes. Minneapolis, MN: Univocal Pub.

Foucault, Michel. 2003. *Society Must Be Defended: Lectures at the Collège de France, 1975–76*. Translated by David Macey. New York: Picador.

———. 2010. *The Birth of Biopolitics: Lectures at the Collège de France, 1978–79*. Edited by Michel Senellart. Translated by Graham Burchell. New York: Palgrave Macmillan.

Frazer, James George. 1996. *The Golden Bough: A Study in Magic and Religion*. Harmondsworth, UK: Penguin Books.

Freeman, Elizabeth. 2010. *Time Binds: Queer Temporalities, Queer Histories*. Durham, NC: Duke University Press.

Freeman, Scott. 2017. "Sovereignty and Soil: Collective and Wage Labor in Rural Haiti." In *Who Owns Haiti? Power, Power, and Sovereignty*, edited by Robert Maguire and Scott Freeman. Gainesville: University Press of Florida.

Friedman, Lisa. 2017. "As U.S. Sheds Role as Climate Change Leader, Who Will Fill the Void?" *New York Times*, November 12, 2017, sec. A.

Ghosh, Amitav. 2017. *The Great Derangement: Climate Change and the Unthinkable*. Chicago: The University of Chicago Press.

Ginsburg, Faye. 1995. "The Parallax Effect: The Impact of Aboriginal Media on Ethnographic Film." *Visual Anthropology Review* 11 (2): 64–76. https://doi.org/10.1525/var.1995.11.2.64.

Gladstone, Jason. 2019. "Lines in the Dirt (c. 1969): Postwar Literalism and the Failure of Technology." *American Literature* 91 (2): 385–416. https://doi.org/10.1215/00029831-7529191.

Goldstein, Ruth. 2019. "Ethnobotanies of Refusal: Methodologies in Respecting Plant(Ed)-human Resistance." *Anthropology Today* 35 (2): 18–22. https://doi.org/10.1111/1467-8322.12495.

Gordillo, Gastón. 2014. *Rubble: The Afterlife of Destruction*. Durham, NC: Duke University Press.

———. 2018. "Terrain as Insurgent Weapon: An Affective Geometry of Warfare in the Mountains of Afghanistan." *Political Geography* 64 (May): 53–62. https://doi.org/10.1016/j.polgeo.2018.03.001.

Goudie, A. S., and N. J. Middleton. 1992. "The Changing Frequency of Dust Storms through Time." *Climatic Change* 20 (3): 197–225. https://doi.org/10 .1007/BF00139839.

Graeter, Stefanie. 2020. "Infrastructural Incorporations: Toxic Storage, Corporate Indemnity, and Ethical Deferral in Peru's Neoextractive Era." *American Anthropologist* 122 (1): 21–36. https://doi.org/10.1111/aman.13367.

Greenhalgh, Susan. 2008. *Just One Child: Science and Policy in Deng's China.* Berkeley: University of California Press.

Guarasci, Bridget. 2015. "The National Park: Reviving Eden in Iraq's Marshes." *Arab Studies Journal* 23 (1): 128–53.

Günel, Gökçe. 2016. "What Is Carbon Dioxide? When Is Carbon Dioxide?" *PoLAR: Political and Legal Anthropology Review* 39 (1): 33–45. https://doi .org/10.1111/plar.12129.

———. 2019. *Spaceship in the Desert: Energy, Climate Change, and Urban Design in Abu Dhabi.* Durham, NC: Duke University Press.

Guo, Jianping, Mengyun Lou, Yucong Miao, Yuan Wang, Zhaoliang Zeng, Huan Liu, Jing He, et al. 2017. "Trans-Pacific Transport of Dust Aerosols from East Asia: Insights Gained from Multiple Observations and Modeling." *Environmental Pollution* 230 (November): 1030–39. https://doi.org/10.1016/j.envpol .2017.07.062.

Guyer, Jane I. 2007. "Prophecy and the Near Future: Thoughts on Macroeconomic, Evangelical, and Punctuated Time." *American Ethnologist* 34 (3): 409–21. https://doi.org/10.1525/ae.2007.34.3.409.

Hacking, Ian. 2002. *Historical Ontology.* Cambridge, MA: Harvard University Press.

Hao, Jingfang. 2015. "Folding Beijing." Translated by Ken Liu. *Uncanny Magazine.* 2015. https://uncannymagazine.com/article/folding-beijing-2/.

Haraway, Donna J. 1988. "Situated Knowledges: The Science Question in Feminism and the Privilege of Partial Perspective." *Feminist Studies* 14 (3): 575. https://doi.org/10.2307/3178066.

———. 1991. *Simians, Cyborgs, and Women: The Reinvention of Nature.* New York: Routledge.

———. 2016. *Staying with the Trouble: Making Kin in the Chthulucene.* Durham, NC: Duke University Press. https://doi.org/10.1215/9780822373780.

Harvey, David. 2011. *A Brief History of Neoliberalism.* Oxford: Oxford University Press.

Hathaway, Michael J. 2013. *Environmental Winds: Making the Global in Southwest China.* Berkeley: University of California Press.

Hecht, Gabrielle. 2018. "Interscalar Vehicles for an African Anthropocene: On Waste, Temporality, and Violence." *Cultural Anthropology* 33 (1): 109–41. https://doi.org/10.14506/ca33.1.05.

Hegel, Georg Wilhelm. 2001. *The Philosophy of History.* Translated by J. Sibree. Kitchener, UK: Batoche Books.

Heilmann, Sebastian, and Elizabeth J. Perry, eds. 2011. "Embracing Uncertainty: Guerrilla Policy Style and Adaptive Governance in China." In *Mao's Invisible Hand: The Political Foundations of Adaptive Governance in China*, 1–29. Harvard Contemporary China Series 17. Cambridge, MA: Harvard University Asia Center : Distributed by Harvard University Press.

Helmreich, Stefan. 2011. "Nature/Culture/Seawater." *American Anthropologist* 113 (1): 132–44. https://doi.org/10.1111/j.1548-1433.2010.01311.x.

Hetherington, Kregg, ed. 2019. *Infrastructure, Environment, and Life in the Anthropocene*. Durham, NC: Duke University Press.

Hillenbrand, Margaret. 2020. *Negative Exposures: Knowing What Not to Know in Contemporary China*. Durham, NC: Duke University Press.

Hirschkind, Charles. 2009. *The Ethical Soundscape: Cassette Sermons and Islamic Counterpublics*. New York: Columbia University Press.

Hoffman, Lisa. 2011. "Urban Modeling and Contemporary Technologies of City-Building in China: The Production of Regimes of Green Urbanism." In *Worlding Cities: Asian Experiments and the Art of Being Global*, edited by Ananya Roy and Aihwa Ong, 55–76. Chichester, UK: Wiley-Blackwell.

Hong, Jiang. 2015. "Taking Down the 'Great Green Wall': The Science and Policy Discourses of Desertification and Its Control in China." In *The End of Desertification? Disputing Environmental Change in the Drylands*, 513–38. New York: Springer-Verlag.

Howe, Cymene. 2015. "Life Above Earth: An Introduction." *Cultural Anthropology* 30 (2): 203–9. https://doi.org/10.14506/ca30.2.03.

———. 2019. *Ecologics: Wind and Power in the Anthropocene*. Durham, NC: Duke University Press.

Huang, Echo. 2016. "China's Giant Smog-Sucking Tower Was Simply No Match for Its Air Pollution." *Quartz*, November 25, 2016. https://qz.com/846093/chinas -giant-smog-sucking-tower-was-simply-no-match-for-its-air-pollution/.

Huang, Shan. 2020. "Space of Care in the City: Doing Fieldwork in the Midst of Hong Kong's Protest Movements in 2019." *Cultural Anthropology Online: Fieldsights* (blog). November 24, 2020. https://culanth.org/fieldsights/space -of-care-in-the-city-doing-fieldwork-in-the-midst-of-hong-kongs-protest -movements-in-2019 .

Huebert, Barry J. 2003. "An Overview of ACE-Asia: Strategies for Quantifying the Relationships between Asian Aerosols and Their Climatic Impacts." *Journal of Geophysical Research* 108 (D23): 8633. https://doi.org/10.1029 /2003JD003550.

Huo, Weiya. 2011. "China's Great Disappearing Lake." *China Dialogue (Zhongguo Duihua)*, January. https://www.chinadialogue.net/article/show/single /en/4068-China-s-great-disappearing-lake.

Ingold, Timothy. 2010. "Footprints through the Weather-World: Walking, Breathing, Knowing." *Journal of the Royal Anthropological Institute* 16 (May): S121–39. https://doi.org/10.1111/j.1467-9655.2010.01613.x.

Jacobs, Andrew. 2011. "The Privileges of China's Elite Include Purified Air." *New York Times*, November 4, 2011.

Jaffe, Dan, Julie Snow, and Owen Cooper. 2003. "The 2001 Asian Dust Events: Transport and Impact on Surface Aerosol Concentrations in the U.S." *Eos, Transactions American Geophysical Union* 84 (46): 501–7. https://doi.org/10.1029/2003EO460001.

Jameson, Frederic. 1992. *Postmodernism; or, The Cultural Logic of Late Capitalism*. Durham, NC: Duke University Press. https://doi.org/10.1215/9780822378419.

———. 2003. "Future City." *New Left Review* 21: 65–79.

Jensen, Casper Bruun, and Atsuro Morita. 2017. "Introduction: Infrastructures as Ontological Experiments." *Ethnos* 82 (4): 615–26. https://doi.org/10.1080/00141844.2015.1107607.

Jeux olympiques d'été, ed. 2010. *Official Report of the Beijing 2008 Olympic Games*. Beijing: BOCOG.

Ji, Dongsheng, Liang Li, Yuesi Wang, Junke Zhang, Mengtian Cheng, Yang Sun, Zirui Liu, et al. 2014. "The Heaviest Particulate Air-Pollution Episodes Occurred in Northern China in January, 2013: Insights Gained from Observation." *Atmospheric Environment* 92 (August): 546–56. https://doi.org/10.1016/j.atmosenv.2014.04.048.

Jin, Feng. 1998. "Jing, the Concept of Scenery in Texts on the Traditional Chinese Garden: An Initial Exploration." *Studies in the History of Gardens & Designed Landscapes* 18 (4): 339–65. https://doi.org/10.1080/14601176.1998.10435557.

Jobson, Ryan Cecil. 2020. "The Case for Letting Anthropology Burn: Sociocultural Anthropology in 2019." *American Anthropologist* 122 (2): 259–71. https://doi.org/10.1111/aman.13398.

Johnson, Alix. 2019. "Data Centers as Infrastructural In-Betweens: Expanding Connections and Enduring Marginalities in Iceland." *American Ethnologist* 46 (1): 75–88. https://doi.org/10.1111/amet.12735.

Johnson, Thomas, and Kathinka Fürst. 2020. "Praying for Blue Skies: Artistic Representations of Air Pollution in China." *Modern China*, November, 009770042096728. https://doi.org/10.1177/0097700420967288.

Jullien, François. 1999. *The Propensity of Things: Toward a History of Efficacy in China*. New York: Zone Books.

———. 2004. *In Praise of Blandness: Proceeding from Chinese Thought and Aesthetics*. New York: Zone Books.

Jung, Min-ho. 2015. "Greenpeace Spares China from Blame for Fine Dust." *Korea Times*, March 4, 2015. https://m.koreatimes.co.kr/pages/article.asp?newsIdx=174608.

Kaganovsky, Lilya. 2009. "The Cultural Logic of Late Socialism." *Studies in Russian and Soviet Cinema* 3 (2): 185–99. https://doi.org/10.1386/srsc.3.2.185_1.

Kaplan, Caren. 2018. *Aerial Aftermaths: Wartime from Above*. Next Wave. Durham, NC: Duke University Press.

Kar, Amal, and Kazuhiko Takeuchi. 2004. "Yellow Dust: An Overview of Research and Felt Needs." *Journal of Arid Environments* 59 (1): 167–87. https://doi.org/10.1016/j.jaridenv.2004.01.010.

Karatani, Kōjin. 2005. *Transcritique: On Kant and Marx*. Cambridge, MA: MIT Press.

Karl, Rebecca E. 2020. *China's Revolutions in the Modern World: A Brief Interpretive History*. London: Verso.

Kay, Samuel. 2020. "Breathing in Beijing: Governing People and Particles in Urban China." In *Disastrous Times: Beyond Environmental Crisis in Urbanizing Asia*, edited by Eli Elinoff and Tyson Vaughan, 25–45. Philadelphia: University of Pennsylvania Press. https://doi.org/10.2307/j.ctv16qjxvp.

Kelly, Ann H., and Javier Lezaun. 2014. "Urban Mosquitoes, Situational Publics, and the Pursuit of Interspecies Separation in Dar Es Salaam: Interspecies Separation in Dar Es Salaam." *American Ethnologist* 41 (2): 368–83. https://doi.org/10.1111/amet.12081.

Ke-Yi, Chen. 2010. "The Northern Path of Asian Dust Transport from the Gobi Desert to North America." *Atmospheric and Oceanic Science Letters* 3 (3): 155–59. https://doi.org/10.1080/16742834.2010.11446858.

Khasbagan, and Soyolt. 2008. "Indigenous Knowledge for Plant Species Diversity: A Case Study of Wild Plants' Folk Names Used by the Mongolians in Ejina Desert Area, Inner Mongolia, P. R. China." *Journal of Ethnobiology and Ethnomedicine* 4 (1): 2. https://doi.org/10.1186/1746-4269-4-2.

Kipnis, Andrew. 2007. "Neoliberalism Reified: Suzhi Discourse and Tropes of Neoliberalism in the People's Republic of China." *Journal of the Royal Anthropological Institute* 13 (2): 383–400. https://doi.org/10.1111/j.1467-9655.2007.00432.x.

Kirksey, Eben. 2015. *Emergent Ecologies*. Durham, NC: Duke University Press.

Klein, Joe. 2019. "Earth Displacements: Draining the Swamp in Sulawesi." Presented at the Annual Meetings of the American Anthropological Association, Vancouver, BC, September 21.

Klein, Naomi. 2008. *The Shock Doctrine: The Rise of Disaster Capitalism*. New York: Picador/H. Holt and Co.

Klinger, Julie Michelle. 2017. *Rare Earth Frontiers: From Terrestrial Subsoils to Lunar Landscapes*. Ithaca, NY: Cornell University Press.

Knopf, Brigitte, and Jiang Kejun. 2017. "Germany and China Take the Lead." *Science* 358 (6363): 569–569. https://doi.org/10.1126/science.aar2525.

Kohn, Eduardo. 2013. *How Forests Think: Toward an Anthropology Beyond the Human*. Berkeley: University of California Press.

Kohrman, Matthew. 2021. "Filtered Life: Air Purification, Gender, and Cigarettes in the People's Republic of China." *Public Culture* (2021) 33 (2 (94)): 161–91.

Koolhaas, Rem. 2004. "Beijing Manifesto." *Wired*, August 2004. www.wired
.com/wired/ archive/12.08/images/FF_120_ beijing.pdf.
Korea Herald. 2015. "Editorial: Yellow Dust Worsens." *The Korea Herald*, Febru-
ary 25, 2015. http://www.koreaherald.com/view.php?ud=20150225000699.
Kosek, Jake. 2006. *Understories: The Political Life of Forests in Northern New
Mexico*. Durham, NC: Duke University Press.
Koselleck, Reinhart. 2004. *Futures Past: On the Semantics of Historical Time*.
New York: Columbia University Press.
Kuriyama, Shigehisa. 1994. "The Imagination of Winds and the Development of
the Chinese Conception of the Body." In *Body, Subject and Power in China*,
edited by Angela Zito and Tani E. Barlow, 23–41. Chicago: University of
Chicago Press.
———. 2011. *The Expressiveness of the Body and the Divergence of Greek and
Chinese Medicine*. New York: Zone Books.
Kwon, Byung Hyun. 2012. "The Moral Equivalent of War: Joining with Our
Chinese Neighbors to Stop the Spread of Deserts in Northeast Asia." Ecocity
Media. November 1, 2012. https://ecocity.wordpress.com/2012/11/01/the
-moral-equivalent-of-war-joining-with-our-chinese-neighbors-to-stop-the
-spread-of-deserts-in-northeast-asia-continued/.
Lao Tzu. 2019. *Tao Te Ching: A Book about the Way and the Power of The Way*.
Translated by Ursula K. Le Guin. Boulder, CO: Shambhala.
Larkin, Brian. 2008. *Signal and Noise: Media, Infrastructure, and Urban
Culture in Nigeria*. Durham, NC: Duke University Press.
———. 2013. "The Politics and Poetics of Infrastructure." *Annual Review of
Anthropology* 42 (1): 327–43. https://doi.org/10.1146/annurev-anthro-092412
-155522.
Latour, Bruno. 2005. *Reassembling the Social: An Introduction to Actor-Network-
Theory*. Oxford: Oxford University Press.
———. 2017. *Facing Gaia: Eight Lectures on the New Climatic Regime*. Trans-
lated by Catherine Porter. Cambridge, UK: Polity.
Lavery, Trish J., Ben Roudnew, Peter Gill, Justin Seymour, Laurent Seuront,
Genevieve Johnson, James G. Mitchell, and Victor Smetacek. 2010. "Iron
Defecation by Sperm Whales Stimulates Carbon Export in the Southern
Ocean." *Proceedings of the Royal Society B: Biological Sciences* 277 (1699):
3527–31. https://doi.org/10.1098/rspb.2010.0863.
Lazier, Benjamin. 2011. "Earthrise; or, The Globalization of the World Picture."
The American Historical Review 116 (3): 602–30. https://doi.org/10.1086/ahr
.116.3.602.
Lee, Li-Young. 1990. *The City in Which I Love You: Poems*. Brockport, NY: BOA
Editions.
Lee, Shannon. 2020. *Be Water, My Friend: The Teachings of Bruce Lee*. New
York: Flatiron Books.

Levenson, Joseph. 1968. *Confucian China and Its Modern Fate: A Trilogy.* Berkeley: University of California Press.

Lévi-Strauss, Claude. 1961. *Tristes Tropiques.* Translated by John Russell. New York: Criterion Books.

———. 2000. *The Savage Mind.* Chicago: University of Chicago Press.

Li, Yifei, and Judith Shapiro. 2020. *China Goes Green: Coercive Environmentalism for a Troubled Planet.* Cambridge, UK : Polity.

Li, Yiyun. 2017. *Dear Friend, From My Life I Write to You in Your Life.* New York: Random House.

Li, YuLin, Chen Jing, Wei Mao, Duo Cui, XinYuan Wang, and XueYong Zhao. 2014. "N and P Resorption in a Pioneer Shrub (*Artemisia halodendron*) Inhabiting Severely Desertified Lands of Northern China." *Journal of Arid Land* 6 (2): 174–85. https://doi.org/10.1007/s40333-013-0222-7.

Li, Zhiming, Huinuan Lin, Long Gu, Jingwen Gao, and Chi-Meng Tzeng. 2016. "Herba Cistanche (Rou Cong-Rong): One of the Best Pharmaceutical Gifts of Traditional Chinese Medicine." *Frontiers in Pharmacology* 7 (March). https://doi.org/10.3389/fphar.2016.00041.

Liebman, Adam. 2019. "Reconfiguring Chinese Natures: Frugality and Waste Reutilization in Mao Era Urban China." *Critical Asian Studies* 51 (4): 537–57. https://doi.org/10.1080/14672715.2019.1658211.

Lin, Jintai, Da Pan, Steven J. Davis, Qiang Zhang, Kebin He, Can Wang, David G. Streets, Donald J. Wuebbles, and Dabo Guan. 2014. "China's International Trade and Air Pollution in the United States." *Proceedings of the National Academy of Sciences* 111 (5): 1736–41. https://doi.org/10.1073/pnas.1312860111.

Liu, Cixin. 2014. *The Three-Body Problem.* Translated by Ken Liu. New York: Tor Books.

Liu, Jianguo, and Jared Diamond. 2005. "China's Environment in a Globalizing World." *Nature* 435 (7046): 1179–86. https://doi.org/10.1038/4351179a.

Liu, Ken. 2014. "Translator's Postscript." In *The Three-Body Problem*, by Cixin Liu, 397–99. New York: Tor Books.

Liu, Petrus. 2015. *Queer Marxism in Two Chinas.* Durham, NC: Duke University Press.

Liu, Xin. 2000. *In One's Own Shadow: An Ethnographic Account of the Condition of Post-Reform Rural China.* Berkeley: University of California Press.

———. 2009. *The Mirage of China: Anti-Humanism, Narcissism, and Corporeality of the Contemporary World.* New York: Berghahn Books.

Livingston, Julie. 2019. *Self-Devouring Growth: A Planetary Parable as Told from Southern Africa.* Durham, NC: Duke University Press.

Logan, D. C., and G. R. Stewart. 1992. "Germination of the Seeds of Parasitic Angiosperms." *Seed Science Research* 2 (4): 179–90. https://doi.org/10.1017/S0960258500001367.

Long, Jane, James Anderson, Ken Caldeira, Joe Chaisson, Steven Hamburg, David Keith, Ron Lehman, et al. 2011. "Geoengineering: A National Strategic Plan for Research on the Potential Effectiveness, Feasibility, and Consequences of Climate Remediation Technologies." Task Force on Climate Remediation Research, The Bipartisan Policy Center. https://keith.seas .harvard.edu/files/tkg/files/bpc_climate_remediation_tf_final_report.pdf.

Lu, Duanfang. 2011. *Remaking Chinese Urban Form: Modernity, Scarcity and Space, 1949–2005*. Abingdon, UK: Routledge.

Luhmann, Niklas. 1998. *Observations on Modernity*. Stanford, CA: Stanford University Press.

Mahajan, Nidhi. 2019. "Dhow Itineraries: The Making of a Shadow Economy in the Western Indian Ocean." *Comparative Studies of South Asia, Africa and the Middle East* 39 (3): 407–19. https://doi.org/10.1215/1089201X-7885356.

Maldonado, Maria T., Szymon Surma, and Evgeny A. Pakhomov. 2016. "Southern Ocean Biological Iron Cycling in the Pre-Whaling and Present Ecosystems." *Philosophical Transactions of the Royal Society A: Mathematical, Physical and Engineering Sciences* 374 (2081). https://doi.org/10.1098/rsta .2015.0292.

Malkki, Liisa. 1992. "National Geographic: The Rooting of Peoples and the Territorialization of National Identity among Scholars and Refugees." *Cultural Anthropology* 7 (1): 24–44. https://doi.org/10.1525/can.1992.7.1.02a00030.

Mao, Zedong. 1945. "The Foolish Old Man Who Removed the Mountains." Marxists.org. https://www.marxists.org/reference/archive/mao/selected -works/volume-3/mswv3_26.htm.

———. 1977. *Selected Works of Mao Tse-Tung, Volume 5*. Oxford: Pergamon Press.

———. 2017. *On Practice and Contradiction*. Edited by Slavoj Žižek. London: Verso.

Marcus, George E. 1995. "Ethnography in/of the World System: The Emergence of Multi-sited Ethnography." *Annual Review of Anthropology* 24 (1): 95–117. https://doi.org/10.1146/annurev.an.24.100195.000523.

Marder, Michael. 2016. *Dust*. Object Lessons. New York: Bloomsbury Academic.

Marx, Karl, and Friedrich Engels. 1979. *Manifesto of the Communist Party*. New York: International Publishers.

Masatoshi, Yoshino. 2002. "Climatology of Yellow Sand (Asian Sand, Asian Dust or Kosa) in East Asia." *Science in China Series D: Earth Sciences* 45 (S1): 59–70. https://doi.org/10.1007/BF02878390.

Masco, Joseph. 2006. *The Nuclear Borderlands: The Manhattan Project in Post-Cold War New Mexico*. Princeton, NJ: Princeton University Press.

———. 2008. "'Survival Is Your Business': Engineering Ruins and Affect in Nuclear America." *Cultural Anthropology* 23 (2): 361–98. https://doi.org/10 .1111/j.1548-1360.2008.00012.x.

———. 2010. "Bad Weather: On Planetary Crisis." *Social Studies of Science* 40 (1): 7–40. https://doi.org/10.1177/0306312709341598.

———. 2014. *The Theater of Operations: National Security Affect from the Cold War to the War on Terror*. Durham: Duke University Press.

Massumi, Brian. 2009. "National Enterprise Emergency: Steps toward an Ecology of Powers." *Theory, Culture & Society* 26 (6): 153–85. https://doi.org /10.1177/0263276409347696.

Mathews, Andrew S. 2017. "Ghostly Forms and Forest Histories." In *Arts of Living on a Damaged Planet: Ghosts and Monsters of the Anthropocene*, edited by Anna Lowenhaupt Tsing, Heather Swanson, Elaine Gan, and Elaine Bubandt. Minneapolis: University of Minnesota Press.

———. 2018. "Landscapes and Throughscapes in Italian Forest Worlds: Thinking Dramatically about the Anthropocene." *Cultural Anthropology* 33 (3): 386–414. https://doi.org/10.14506/ca33.3.05.

Mauss, Marcel. 2007. "Techniques of the Body." In *Beyond the Body Proper: Reading the Anthropology of Material Life*, edited by Margaret M. Lock and Judith Farquhar, 50–68. Body, Commodity, Text. Durham, NC: Duke University Press.

Mbembe, Achille. 2003. "Necropolitics." Translated by Libby Meintjes. *Public Culture* 15 (1): 11–40.

McCormack, Derek P. 2018. *Atmospheric Things: On the Allure of Elemental Envelopment*. Durham, NC: Duke University Press.

Mertha, Andrew. 2009. "'Fragmented Authoritarianism 2.0': Political Pluralization in the Chinese Policy Process." *The China Quarterly* 200 (December): 995–1012. https://doi.org/10.1017/S0305741009990592.

Mettrie, Julien Offray de. 1996. *Machine Man and Other Writings*. Edited by Ann Thomson. Cambridge: Cambridge University Press. https://doi.org/10 .1017/CBO9781139166713.

Miéville, China. 2009. *The City & The City*. New York: Del Rey Ballantine Books.

Mitman, Gregg, Michelle Murphy, and Christopher Sellers. 2004. "Introduction: A Cloud over History." In *Landscapes of Exposure: Knowledge and Illness in Modern Environments*, edited by Gregg Mitman, Michelle Murphy, and Christopher Sellers. *Osiris* 19. Chicago: University of Chicago Press.

Miyazaki, Hirokazu. 2004. *The Method of Hope: Anthropology, Philosophy, and Fijian Knowledge*. Stanford, CA: Stanford University Press.

Mol, Annemarie. 2002. *The Body Multiple: Ontology in Medical Practice*. Durham, NC: Duke University Press.

Moore, Jason W. 2015. *Capitalism in the Web of Life: Ecology and the Accumulation of Capital*. New York: Verso.

Murphy, Michelle. 2006. *Sick Building Syndrome and the Problem of Uncertainty: Environmental Politics, Technoscience, and Women Workers*. Durham, NC: Duke University Press.

———. 2017. "Alterlife and Decolonial Chemical Relations." *Cultural Anthropology* 32 (4): 494–503. https://doi.org/10.14506/ca32.4.02.

Myers, Natasha. 2015. *Rendering Life Molecular: Models, Modelers, and Excitable Matter*. Durham, NC: Duke University Press.

NASA Earth Observatory. 2007. "A Perfect (Dust) Storm." *NASA Earth Observatory* (blog). December 26, 2007. https://earthobservatory.nasa.gov/images/8341/a-perfect-dust-storm.

Neyrat, Frédéric. 2018. *Atopias: Manifesto for a Radical Existentialism*. New York: Fordham University Press.

———. 2019. *The Unconstructable Earth: An Ecology of Separation*. New York: Fordham University Press.

Ng, Emily. 2020. *A Time of Lost Gods: Mediumship, Madness, and the Ghost after Mao*. Oakland: University of California Press.

Nicol, Stephen, Andrew Bowie, Simon Jarman, Delphine Lannuzel, Klaus M. Meiners, and Pier Van Der Merwe. 2010. "Southern Ocean Iron Fertilization by Baleen Whales and Antarctic Krill: Whales, Antarctic Krill and Iron Fertilization." *Fish and Fisheries* 11 (2): 203–9. https://doi.org/10.1111/j.1467-2979.2010.00356.x.

Nixon, Robert. 2013. *Slow Violence and the Environmentalism of the Poor*. Cambridge, MA: Harvard University Press.

Nye, David E. 1994. *American Technological Sublime*. Cambridge, MA: MIT Press.

O'Neill, Bruce. 2017. *The Space of Boredom: Homelessness in the Slowing Global Order*. Durham, NC: Duke University Press.

Ong, Aihwa. 2006. *Neoliberalism as Exception: Mutations in Citizenship and Sovereignty*. Durham, NC: Duke University Press.

———. 2011. "Hyperbuilding: Spectacle, Speculation, and the Hyperspace of Sovereignty." In *Worlding Cities: Asian Experiments and the Art of Being Global*, edited by Ananya Roy and Aihwa Ong, 205–26. Oxford, UK: Wiley-Blackwell. https://doi.org/10.1002/9781444346800.

———. 2016. *Fungible Life: Experiment in the Asian City of Life*. Durham, NC: Duke University Press.

Ong, Aihwa, and Li Zhang. 2008. "Privatizing China: Powers of Self, Socialism from Afar." In *Privatizing China: Socialism from Afar*, edited by Li Zhang and Aihwa Ong, 1–20. Ithaca: Cornell University Press.

Oreskes, Naomi, and Erik M. Conway. 2014. *The Collapse of Western Civilization: A View from the Future*. New York: Columbia University Press.

Ortiz, Edward. 2015. "Pollution from China Taking High Route to Central Valley, Study Says." *Sacramento Bee*, March 31, 2015, sec. Environment. https://www.sacbee.com/news/local/environment/article17057207.html.

Palmer, James. 2017. "Dispatch: How to Destroy the Heart of a Chinese City." *Foreign Policy*. https://foreignpolicy.com/2017/05/31/how-to-destroy-the-heart-of-a-chinese-city-beijing/.

Pan, Yue. 2007. "On Socialist Ecological Civilization (Lun Shehuizhuyi Shengtai Wenming)." *Green Leaf (Lüye)*, February 6, 2007. https://wenku.baidu.com/view/acd66d515901020207409c8d.html.

Pandian, Anand. 2018. "Ursula K. Le Guin, Interplanetary Anthropologist."
  Anthro{dendum}. January 26, 2018. https://anthrodendum.org/2018/01/26
  /ursula-ak-le-guin-interplanetary-anthropologist-2/.
———. 2019. *A Possible Anthropology: Methods for Uneasy Times*. Durham, NC:
  Duke University Press.
Parikka, Jussi. 2015. *A Geology of Media*. Minneapolis: University of Minnesota
  Press.
Park, Sangho. 2011. "The Seeds of Hope: Combating Desertification of Inner
  Mongolia and Preserving the Steppe." Seoul: Ecopeace Asia & Hyundai
  Motor Company.
Pershing, Andrew J., Line B. Christensen, Nicholas R. Record, Graham D.
  Sherwood, and Peter B. Stetson. 2010. "The Impact of Whaling on the Ocean
  Carbon Cycle: Why Bigger Was Better." Edited by Stuart Humphries. *PLoS
  ONE* 5 (8): e12444. https://doi.org/10.1371/journal.pone.0012444.
Petryna, Adriana. 2018. "Wildfires at the Edges of Science: Horizoning Work
  amid Runaway Change." *Cultural Anthropology* 33 (4): 570–95. https://doi
  .org/10.14506/ca33.4.06.
Philip, Kavita. 2004. *Civilizing Natures: Race, Resources, and Modernity in
  Colonial South India*. New Brunswick, N.J: Rutgers University Press.
Phillips, Tom. 2017. "China's Xi Jinping Says Paris Climate Deal Must Not Be
  Allowed to Fail." *The Guardian*, January 18, 2017. https://www.theguardian
  .com/world/2017/jan/19/chinas-xi-jinping-says-world-must-implement-paris
  -climate-deal.
Pieke, Frank N. 2014. "Anthropology, China, and the Chinese Century." *Annual
  Review of Anthropology* 43: 123–38.
Pine, Jason. 2019. *The Alchemy of Meth: A Decomposition*. Minneapolis:
  University of Minnesota Press.
Pinkus, Karen, and Derek Woods. 2019. "From the Editors: Terraforming."
  Diacritics 47 (3): 4–5. https://doi.org/10.1353/dia.2019.0023.
Povinelli, Elizabeth A. 2008. "The Child in the Broom Closet: States of Killing
  and Letting Die." *South Atlantic Quarterly* 107 (3): 509–30. https://doi.org/10
  .1215/00382876-2008-004.
———. 2016. *Geontologies: A Requiem to Late Liberalism*. Durham, NC: Duke
  University Press.
———. 2017. "Fires, Fogs, Winds." *Cultural Anthropology* 32 (4): 504–13. https://
  doi.org/10.14506/ca32.4.03.
Pyne, Stephen J. 2015. *Between Two Fires: A Fire History of Contemporary
  America*. Tucson: University of Arizona Press.
Qian, Xuesen. 2012. *Readings on the Theory of the Sixth Industrial Revolution
  (Di Liuci Chanye Geming Lilun Xuexi Duben)*. Edited by Xi'an Jiaotong
  Daxue. Xi'an: Xi'an Jiaotong Daxue Chuban She.
Qiu, Jane, and Daniel Cressley. 2008. "Meteorology: Taming the Sky." *Nature*
  453: 970–74.

Rabinow, Paul, and Gaymon Bennett. 2008. "From Bio-Ethics to Human Practices; or, Assembling Contemporary Equipment." In *Tactical Biopolitics: Art, Activism, and Technoscience*, edited by Beatriz da Costa and Kavita Philip, 389–201. Cambridge, MA: MIT Press.

Raffles, Hugh. 2002. *In Amazonia: A Natural History*. Princeton, NJ: Princeton University Press.

———. 2011. *Insectopedia*. New York: Vintage.

———. 2013. "6 Stones to Get Lost With." *Orion Magazine*, February 27, 2013. https://orionmagazine.org/article/6-stones-to-get-lost-with/.

———. 2020. *The Book of Unconformities: Speculations on Lost Time*. New York: Pantheon Books.

Ramzy, Austin. 2009. "Twittering Beijing's Bad Air." *Time*, June 20, 2009. https://world.time.com/2009/06/20/twittering-beijings-bad-air/.

Rappaport, Roy A. 1984. *Pigs for the Ancestors: Ritual in the Ecology of a New Guinea People*. New Haven, CT: Yale University Press.

Reddy, Elizabeth. 2014. "What Does It Mean to Do Anthropology in the Anthropocene?" Platypus: The CASTAC Blog. 2014. http://blog.castac.org/2014/04/what-does-it-mean-to-do-anthropology-in-the-anthropocene/.

Renminbao (People's Daily). 2000. "Zhurongji Xialing 'Shayang Hucao' Bao Beijing. (Zhu Rongji Commands 'Kill Goats and Protect Grass' to Protect Beijing)." *Renminbao (People's Daily)*, August 23, 2000. http://www.renminbao.com/rmb/articles/2000/8/23/2504.html.

Rhee, Margaret. 2015. "Racial Recalibration." *Asian Diasporic Visual Cultures and the Americas* 1 (3): 285–309. https://doi.org/10.1163/23523085-00103004.

Rheinberger, Hans-Jörg. 1995. "From Experimental Systems to Cultures of Experimentation." In *Concepts, Theories, and Rationality in the Biological Sciences*, edited by Gereon Wolters and James G. Lennox, 107–22. Pittsburgh: University of Pittsburgh Press.

Rivera, Takeo. 2014. "Do Asians Dream of Electric Shrieks? Techno-Orientalism and Erotohistoriographic Masochism in Eidos Montreal's *Deus Ex: Human Revolution*." *Amerasia Journal* 40 (2): 67–87. https://doi.org/10.17953/amer.40.2.j012284wu6230604.

Robinson, Kim Stanley. 2013. *2312*. New York: Orbit.

Rofel, Lisa. 1999. *Other Modernities: Gendered Yearnings in China after Socialism*. Berkeley: University of California Press.

———. 2007. *Desiring China: Experiments in Neoliberalism, Sexuality, and Public Culture*. Durham, NC: Duke University Press.

Rogaski, Ruth. 2019. "Air/*Qi* Connections and China's Smog Crisis: Notes from the History of Science." *Cross-Currents* 30: 55–77.

Rosaldo, Renato. 2014. *The Day of Shelly's Death: The Poetry and Ethnography of Grief*. Durham, NC: Duke University Press.

Sayre, Nathan F. 2008. "The Genesis, History, and Limits of Carrying Capacity." *Annals of the Association of American Geographers* 98 (1): 120–34. https://doi .org/10.1080/00045600701734356.

Schaffer, Simon, and Steven Shapin. 2017. *Leviathan and the Air-Pump: Hobbes, Boyle, and the Experimental Life*. Princeton, NJ: Princeton University Press.

Schein, Louisa. 1997. "Gender and Internal Orientalism in China." *Modern China* 23 (1): 69–98. https://doi.org/10.1177/009770049702300103.

Schmitt, Carl. 2007. *The Concept of the Political*. Translated by George Schwab. Chicago: University of Chicago Press.

Schrader, Ann. 2001. "Latest Import from China: Haze, Asian Dust Blanketing Denver, U.S." *Denver Post*, April 18, 2001, WED1 edition, sec. A.

Schuetze, Christopher F. 2014. "Turning Dust to Diamonds." *New York Times*, March 17, 2014, International Edition.

Schwenk, Theodor. 1965. *Sensitive Chaos: The Creation of Flowing Forms in Water and Air*. London: Rudolf Steiner Press.

Schwenkel, Christina. 2015. "Spectacular Infrastructure and Its Breakdown in Socialist Vietnam." *American Ethnologist* 42 (3): 520–34. https://doi.org/10 .1111/amet.12145.

Scott, David. 2004. *Conscripts of Modernity: The Tragedy of Colonial Enlightenment*. Durham, NC: Duke University Press.

———. 2014. *Omens of Adversity: Tragedy, Time, Memory, Justice*. Durham, NC: Duke University Press.

Scott, James C. 1998. *Seeing Like a State: How Certain Schemes to Improve the Human Condition Have Failed*. New Haven, CT: Yale University Press.

Searle, John R. 1980. "Minds, Brains, and Programs." *Behavioral and Brain Sciences* 3 (3): 417–24. https://doi.org/10.1017/S0140525X00005756.

Serres, Michel. 1995. *Conversations on Science, Culture, and Time: Michel Serres with Bruno Latour*. Translated by Roxanne Lapidus. Ann Arbor: University of Michigan Press. https://doi.org/10.3998/mpub.9736.

———. 2012. *Biogea*. Translated by M. Randolph Burks. Minneapolis, MN: Univocal.

Shange, Savannah. 2019. *Progressive Dystopia: Abolition, Anthropology, and Race in the New San Francisco*. Durham, NC: Duke University Press.

Shapiro, Judith. 2001. *Mao's War Against Nature: Politics and the Environment in Revolutionary China*. Cambridge, UK: Cambridge University Press. https://doi.org/10.1017/CBO9780511512063.

Shapiro, Nicholas. 2015. "Attuning to the Chemosphere: Domestic Formaldehyde, Bodily Reasoning, and the Chemical Sublime." *Cultural Anthropology* 30 (3): 368–93. https://doi.org/10.14506/ca30.3.02.

Shapiro, Nicholas, Nasser Zakariya, and Jody Roberts. 2017. "A Wary Alliance: From Enumerating the Environment to Inviting Apprehension." *Engaging*

Science, Technology, and Society 3 (September): 575. https://doi.org/10.17351
/ests2017.133.

Shen, Grace Yen. 2014. *Unearthing the Nation: Modern Geology and Nationalism in Republican China.* Chicago: University of Chicago Press.

Simmons, Kristen. 2017. "Settler Atmospherics." *Cultural Anthropology Online
Supplement* (blog). November 20, 2017. https://culanth.org/fieldsights/settler
-atmospherics.

Simondon, Gilbert. 2020. *Individuation in Light of Notions of Form and
Information.* Translated by Taylor Adkins. Minneapolis: University of
Minnesota Press.

Singh, Neera M. 2013. "The Affective Labor of Growing Forests and the Becoming of Environmental Subjects: Rethinking Environmentality in Odisha,
India." *Geoforum* 47 (June): 189–98. https://doi.org/10.1016/j.geoforum.2013
.01.010.

Škof, Lenart, and Petri Berndtson, eds. 2018. *Atmospheres of Breathing.* Albany:
State University of New York Press.

Sloterdijk, Peter. 2009. *Terror from the Air.* Translated by Amy Patton and Steve
Corcoran. Los Angeles: Semiotext(e).

———. 2011. *Bubbles.* Los Angeles, CA: Semiotext(e).

———. 2016. *Foams.* Los Angeles, CA: Semiotext(e).

Song, Hoon. 2012. "Two Is Infinite, Gender Is Post-Social in Papua New
Guinea." *Angelaki* 17 (2): 123–44. https://doi.org/10.1080/0969725X.2012
.701053.

Songster, E. Elena. 2003. "Cultivating the Nation in Fujian's Forests: Forest
Policies and Afforestation Efforts in China, 1911–1937." *Environmental
History* 8 (3): 452. https://doi.org/10.2307/3986204.

Spina, Marco. 2019. "Annual Cashmere Market Report, 2018." The Schneider
Group. https://www.gschneider.com/2019/03/01/annual-cashmere-market
-report/.

Spivak, Gayatri Chakravorty. 2008. *Other Asias.* Malden, MA: Blackwell
Publishing.

Star, Susan Leigh. 1999. "The Ethnography of Infrastructure." *American
Behavioral Scientist* 43 (3): 377–91. https://doi.org/10.1177
/00027649921955326.

Star, Susan Leigh, and James R. Griesemer. 1989. "Institutional Ecology,
`Translations' and Boundary Objects: Amateurs and Professionals in
Berkeley's Museum of Vertebrate Zoology, 1907-39." *Social Studies of Science*
19 (3): 387–420. https://doi.org/10.1177/030631289019003001.

State Forestry Administration. 2011. "A Bulletin of Status Quo of Desertification
and Sandification in China (Zhongguo Huangmahua He Shahua Zhuang-
kuang Gongbao)." http://www.forestry.gov.cn/uploadfile/main/2011-1/file
/2011-1-5-59315b03587b4d7793d5d9c3aae7ca86.pdf.

Stein, Susanne. 2015. "Coping with the 'World's Biggest Dust Bowl': Towards a History of China's Forest Shelterbelts, 1950s-Present." *Global Environment* 8 (2): 320–48. https://doi.org/10.3197/ge.2015.080204.

Stengers, Isabelle. 2010. *Cosmopolitics*. Minneapolis: University of Minnesota Press.

Stevens, Wallace. 1954. "Thirteen Ways of Looking at a Blackbird." In *The Collected Poems of Wallace Stevens*. New York: Random House.

Stevenson, Lisa. 2014. *Life Beside Itself: Imagining Care in the Canadian Arctic.* Oakland: University of California Press.

———. 2020. "Looking Away." *Cultural Anthropology* 35 (1). https://doi.org/10.14506/ca35.1.02.

Strathern, Marilyn. 1980. "No Nature, No Culture: The Hagen Case." In *Nature, Culture, and Gender*, edited by Carol P. MacCormack and Marilyn Strathern, 174–222. New York: Cambridge University Press.

———. 1988. *The Gender of the Gift: Problems with Women and Problems with Society in Melanesia.* Berkeley: University of California Press.

———. 1995. *The Relation: Issues in Complexity and Scale.* Cambridge: Prickly Pear Press.

———. 2004. *Partial Connections.* Walnut Creek, Ca.: AltaMira Press.

Sun, Danfeng, Richard Dawson, Hong Li, and Baoguo Li. 2005. "Modeling Desertification Change in Minqin County, China." *Environmental Monitoring and Assessment* 108 (1–3): 169–88. https://doi.org/10.1007/s10661-005-4221-9.

Sun, Yat-Sen. 1927. *San Min Chu I: The Three Principles of the People.* Edited by Li-Ting Chen. Translated by Frank W. Price. Shanghai, China: China Committee, Institute of Foreign Relations.

T and Squatting. 2020. "Situating the Hong Kong Left." *Lausan* (blog). July 30, 2020. https://lausan.hk/2020/situating-the-hong-kong-left/.

Takemura, Toshihiko, Itsushi Uno, Teruyuki Nakajima, Akiko Higurashi, and Itaru Sano. 2002. "Modeling Study of Long-Range Transport of Asian Dust and Anthropogenic Aerosols from East Asia." *Geophysical Research Letters* 29 (24): 11-1-11-4. https://doi.org/10.1029/2002GL016251.

Taylor, Jesse O. 2016. *The Sky of Our Manufacture: The London Fog in British Fiction from Dickens to Woolf.* Charlottesville: University of Virginia Press.

Thongchai Winichakul. 1997. *Siam Mapped: A History of the Geo-Body of a Nation.* Honolulu: University of Hawaii Press.

Tironi, Manuel. 2019. "Lithic Abstractions: Geophysical Operations against the Anthropocene." *Distinktion: Journal of Social Theory* 20 (3): 284–300. https://doi.org/10.1080/1600910X.2019.1610017.

Tomba, Luigi. 2009. "Of Quality, Harmony, and Community: Civilization and the Middle Class in Urban China." *Positions: East Asia Cultures Critique* 17 (3): 591–616. https://doi.org/10.1215/10679847-2009-016.

Tran, Eric. 2020. "Starting with a Line by Joyce Byers." In *The Gutter Spread Guide to Prayer*. Pittsburgh: Autumn House Press.

Tsing, Anna Lowenhaupt. 1993. *In the Realm of the Diamond Queen: Marginality in an Out-of-the-Way Place*. Princeton, NJ: Princeton University Press.

———. 2000. "The Global Situation." *Cultural Anthropology* 15 (3): 327–60. https://doi.org/10.1525/can.2000.15.3.327.

———. 2009. "Supply Chains and the Human Condition." *Rethinking Marxism* 21 (2): 148–76. https://doi.org/10.1080/08935690902743088.

———. 2015. *The Mushroom at the End of the World: On the Possibility of Life in Capitalist Ruins*. Princeton, NJ: Princeton University Press.

Tsing, Anna Lowenhaupt, Andrew S. Mathews, and Nils Bubandt. 2019. "Patchy Anthropocene: Landscape Structure, Multispecies History, and the Retooling of Anthropology: An Introduction to Supplement 20." *Current Anthropology* 60 (S20): S186–97. https://doi.org/10.1086/703391.

Tuan, Yi-Fu. 2001. "The Desert and I: A Study in Affinity." *Michigan Quarterly Review* 40 (1).

US Embassy, Beijing. n.d. "Extremely High Levels of PM2.5: Steps to Reduce Your Exposure." Accessed April 2, 2020. https://china.usembassy-china.org/.cn/embassy-consulates/beijing/air-quality-monitor/extremely-high-levels-pm2-5-steps-reduce-exposure/.

USDA Agricultural Research Service, National Plant Germplasm System. 2020. "*Cistanche deserticola* Ma." 2020. https://npgsweb.ars-grin.gov/gringlobal/taxonomydetail.aspx?id=418363.

Verdery, Katherine. 1996. *What Was Socialism, and What Comes Next?* Princeton, NJ: Princeton University Press.

Virgil. 2007. *The Aeneid of Virgil*. Translated by Allen Mandelbaum. Berkeley, Ca: University of California Press.

Virilio, Paul. 2006. *Speed and Politics*. Los Angeles: Semiotext(e).

Viveiros de Castro, Eduardo. 2012. *Cosmological Perspectivism in Amazonia and Elsewhere: Four Lectures given in the Department of Social Anthropology, Cambridge University, February-March 1998*. Manchester, UK: HAU, Journal of Ethnographic Theory.

Vogel, Ezra F. 2013. *Deng Xiaoping and the Transformation of China*. Cambridge, MA: Harvard University Press.

Vuong, Ocean. 2016. "On Earth We're Briefly Gorgeous." *Night Sky with Exit Wounds*. Port Townsend, Washington: Copper Canyon Press.

Wagner, Roy. 1986. *Symbols That Stand for Themselves*. Chicago: University of Chicago Press.

Wallerstein, Immanuel. 1974. "The Rise and Future Demise of the World Capitalist System: Concepts for Comparative Analysis." *Comparative Studies in Society and History* 16 (4): 387–415.

Wang, Hui. 2011. *The Politics of Imagining Asia*. Translated by Theodore Huters. Cambridge, MA: Harvard University Press.

Wang, Qiang. 2013. "China's Citizens Must Act to Save Their Environment." *Nature* 159 (May): 159.

Wang, Shichuan. 2014. "Experiencing 'Breathing Together, Sharing a Common Destiny' in Xi Jinping's Walk through Nanluoguxiang (Cong Xi Jinping Guang Nanluoguxiang Ganshou 'Tong Huxi, Gong Minyun.'" *Fenghuang Pinlun*, February 25, 2014. http://news.ifeng.com/opinion/wangping/nanluoguxiang/.

Wang, Tao, and Halin Zhao. 2005. "Fifty Years of Desert Research in China [Zhongguo Shamohua Kexue de Wushi Nian]." *Journal of Desert Research* 2: 145–65.

Wang, Tian, Xiaoying Zhang, and Wenyan Xie. 2012. "Cistanche Deserticola Y. C. Ma, 'Desert Ginseng': A Review." *American Journal of Chinese Medicine* 40 (06): 1123–41. https://doi.org/10.1142/S0192415X12500838.

Wang, X., T. Liu, F. Bernard, X. Ding, S. Wen, Y. Zhang, Z. Zhang, et al. 2014. "Design and Characterization of a Smog Chamber for Studying Gas-Phase Chemical Mechanisms and Aerosol Formation." *Atmospheric Measurement Techniques* 7 (1): 301–13. https://doi.org/10.5194/amt-7-301-2014.

Wang, Xunming, Zhibao Dong, Jiawu Zhang, and Lichao Liu. 2004. "Modern Dust Storms in China: An Overview." *Journal of Arid Environments* 58 (4): 559–74. https://doi.org/10.1016/j.jaridenv.2003.11.009.

Wang, Yanan. 2008. "Day for Combating Drought and Desertification: Witnessing Chery's Thousand-Mu Shelterbelt Forest (Fangzhi Shamohua He Ganhan Ri: Jianzheng Qian Mu Qirui Fanghulin)." *Sohu Auto (Souhu Qiche)*, June 17, 2008. http://auto.sohu.com/20080617/n257560919.shtml.

Weiss-Penzias, Peter S., Michael S. Bank, Deana L. Clifford, Alicia Torregrosa, Belle Zheng, Wendy Lin, and Christopher C. Wilmers. 2019. "Marine Fog Inputs Appear to Increase Methylmercury Bioaccumulation in a Coastal Terrestrial Food Web." *Scientific Reports* 9 (1): 17611. https://doi.org/10.1038/s41598-019-54056-7.

Weizman, Eyal. 2017. *Hollow Land: Israel's Architecture of Occupation*. London: Verso.

Welland, Michael. 2009. *Sand: The Never-Ending Story*. Berkeley: University of California Press. http://search.ebscohost.com/login.aspx?direct=true&scope=site&db=nlebk&db=nlabk&AN=672937.

West, Paige. 2006. *Conservation Is Our Government Now: The Politics of Ecology in Papua New Guinea*. Durham, NC: Duke University Press.

White, Hayden V. 2000. *Metahistory: The Historical Imagination in Nineteenth-Century Europe*. Baltimore: Johns Hopkins University Press.

White, Richard. 1995. *The Organic Machine: The Remaking of the Columbia River*. New York: Hill and Wang.

White, Thomas. 2016. "From Sent-Down Youth to Scaled-Up Town." *Inner Asia* 18 (1): 15–36. https://doi.org/10.1163/22105018-12340051.

Whitington, Jerome. 2013. "Fingerprint, Bellwether, Model Event: Climate Change as Speculative Anthropology." *Anthropological Theory* 13 (4): 308–28. https://doi.org/10.1177/1463499613509992.

———. 2016. "What Does Climate Change Demand of Anthropology?" *PoLAR: Political and Legal Anthropology Review* 39 (1): 7–15. https://doi.org/10.1111/plar.12127.

———. 2019. *Anthropogenic Rivers: The Production of Uncertainty in Lao Hydropower.* Ithaca, NY: Cornell University.

———. 2020. "Earth's Data: Climate Change, Thai Carbon Markets, and the Planetary Atmosphere." *American Anthropologist* 122(4): 814–26.

Williams, Dee Mack. 2002. *Beyond Great Walls: Environment, Identity, and Development on the Chinese Grasslands of Inner Mongolia.* Stanford, CA: Stanford University Press.

Wittfogel, Karl. 1957. *Oriental Despotism: A Comparative Study of Total Power.* New Haven, CT: Yale University Press.

Wong, Edward. 2013. "In China, Breathing Becomes a Childhood Risk." *New York Times*, April 22, 2013.

Woods, Derek. 2019. "'Terraforming Earth': Climate and Recursivity." *Diacritics* 47 (3): 6–29. https://doi.org/10.1353/dia.2019.0024.

Worster, Donald. 1990. "The Ecology of Order and Chaos." *Environmental History Review* 14 (1–2): 1–18. https://doi.org/10.2307/3984623.

———. 2004. *Dust Bowl: The Southern Plains in the 1930s.* New York: Oxford University Press.

Wu, Jie, Jianzhou Wang, and Dezhong Chi. 2013. "Wind Energy Potential Assessment for the Site of Inner Mongolia in China." *Renewable and Sustainable Energy Reviews* 21 (May): 215–28. https://doi.org/10.1016/j.rser.2012.12.060.

Wu, Shan, Zifeng Lü, Jiming Hao, Zhe Zhao, Junhua Li, Hideto Takekawa, Hiroaki Minoura, and Akio Yasuda. 2007. "Construction and Characterization of an Atmospheric Simulation Smog Chamber." *Advances in Atmospheric Sciences* 24 (2): 250–58. https://doi.org/10.1007/s00376-007-0250-3.

Xinhua. 2007. "Premier Vows to Curb Desertification." *China Daily*, October 2, 2007. http://www.chinadaily.com.cn/china/2007-10/02/content_6150889.htm.

———. 2016. "Beijing Installs Ventilation Corridors." *Xinhua*, 2016. http://www.xinhuanet.com/english/2016-02/21/c_135116438.htm.

Xinjing Bao. 2014. "Zhongguo Nijian Shijie Zuida Wumai Shiyanshi (China Proposes Construction of the World's Largest Smog Chamber)." *Xinjing Bao (The Beijing News)*, March 2, 2014. bjnews.com.cn/html/2014- 03/02/content_497492. htm?div=0.

Xu, Jintao, Runsheng Yin, Zhou Li, and Can Liu. 2006. "China's Ecological Rehabilitation: Unprecedented Efforts, Dramatic Impacts, and Requisite Policies." *Ecological Economics* 57 (4): 595–607. https://doi.org/10.1016/j.ecolecon.2005.05.008.

Yan, Hairong, and Daniel Vukovich. 2007. "Guest Editor's Introduction: What's Left of Asia." *Positions: East Asia Cultures Critique* 15 (2): 211–24. https://doi.org/10.1215/10679847-2006-029.

Yang, Fan. 2016. "Under the Dome: 'Chinese' Smog as a Viral Media Event." *Critical Studies in Media Communication* 33 (3): 232–44. https://doi.org/10 .1080/15295036.2016.1170172.

Yeh, Emily T. 2005. "Green Governmentality and Pastoralism in Western China: 'Converting Pastures to Grasslands.'" *Nomadic Peoples* 9 (1): 9–30. https://doi.org/10.3167/082279405781826164.

———. 2009. "Greening Western China: A Critical View." *Geoforum* 40 (5): 884–94. https://doi.org/10.1016/j.geoforum.2009.06.004.

Yurchak, Alexei. 1997. "The Cynical Reason of Late Socialism: Power, Pretense, and the Anekdot." *Public Culture* 9: 151–88.

———. 2005. *Everything Was Forever, until It Was No More: The Last Soviet Generation*. Princeton, NJ: Princeton University Press.

Zaloom, Caitlin. 2004. "The Productive Life of Risk." *Cultural Anthropology* 19 (3): 365–91. https://doi.org/10.1525/can.2004.19.3.365.

Zapruder, Matthew. 2012. "Poem for Japan." Poets.Org. 2012. https://poets.org /poem/poem-japan.

Zee, Jerry C. 2015. "States of the Wind: Dust Storms and a Political Meteorology of Contemporary China." Dissertation, UC Berkeley Department of Anthropology. Berkeley, CA.

———. 2017. "Holding Patterns: Sand and Political Time at China's Desert Shores." *Cultural Anthropology* 32 (2): 215–41. https://doi.org/10.14506 /ca32.2.06.

———. 2019. "Groundwork in the Margins: Symbiotic Governance in a Chinese Dust-Shed." In *Frontier Assemblages: The Emergent Politics of Resource Frontiers in Asia*, edited by Jason Cons and Michael Eilenberg, 59–74. Hoboken, NJ: John Wiley & Sons.

———. 2020a. "Asia as Strategy: Deployments of a Chinese Planet." *Hau*.

———. 2020b. "Downwind: Three Phases of an Aerosol Form." In *Voluminous States: Sovereignty, Materiality, and the Territorial Imagination*, edited by Franck Billé, 119–30. Durham, NC: Duke University Press.

———. 2020c. "Mercury Fog Links China's Power Structure to the Deep Time of Coal Formation." In *Feral Atlas: The More-than-Human Anthropocene*, edited by Anna L. Tsing, Jennifer Deger, Alder Keleman Saxena, and Feifei Zhou. Stanford, CA: Stanford University Press Digital. https://feralatlas .supdigital.org/poster/mercury-fog-links-chinas-power-structure-to-the -deep-time-of-coal-formation.

———. 2020d. "Machine Sky: Social and Terrestrial Engineering in a Chinese Weather System." *American Anthropologist* 122 (1): 9–20. https://doi.org/10 .1111/aman.13360.

———. 2020e. "The Dust Kaleidoscope." *Theorizing the Contemporary, Society for Cultural Anthropology Website* (blog). September 22, 2020. https:// culanth.org/fieldsights/the-dust-kaleidoscope.

Zhan, Mei. 2009. *Other-Worldly: Making Chinese Medicine through Trans-national Frames*. Durham, NC: Duke University Press.

———. 2011. "Worlding Oneness: Daoism, Heidegger, and Possibilities for Treating the Human." *Social Text* 29 (4): 107–28. https://doi.org/10.1215/01642472-1416109.

———. 2019. "Out of Nothing: (Re)Worlding 'Theory' through Chinese Medical Entrepreneurship." In *The World Multiple: The Quotidian Politics of Knowing and Generating Entangled Worlds*, edited by Keiichi Ōmura, Grant Otsuki, Shiho Satsuka, and Atsurō Morita, 175–89. New York: Routledge.

Zhang, Amy. 2020. "Circularity and Enclosures: Metabolizing Waste with the Black Soldier Fly." *Cultural Anthropology* 35 (1). https://doi.org/10.14506/ca35.1.08.

Zhang, C. Z., S. X. Wang, Y. Zhang, J. P. Chen, and X. M. Liang. 2005. "In Vitro Estrogenic Activities of Chinese Medicinal Plants Traditionally Used for the Management of Menopausal Symptoms." *Journal of Ethnopharmacology* 98 (3): 295–300. https://doi.org/10.1016/j.jep.2005.01.033.

Zhang, Jiacheng, and Thomas J. Crowley. 1989. "Historical Climate Records in China and Reconstruction of Past Climates." *Journal of Climate* 2 (8): 833–49.

Zhang, Li. 2010. *In Search of Paradise: Middle-Class Living in a Chinese Metropolis*. Ithaca, NY: Cornell University Press.

———. 2020. *Anxious China: Inner Revolution and Politics of Psychotherapy*. Oakland: University of California Press.

Zhang, Xudong. 2001. "The Making of the Post-Tiananmen Intellectual Field: A Critical Overview." In *Whither China: Intellectual Politics in Contemporary China*, edited by Xudong Zhang. Durham, NC: Duke University Press.

Zhao, Jijun, and Jan Woudstra. 2007. "'In Agriculture, Learn from Dazhai': Mao Zedong's Revolutionary Model Village and the Battle against Nature." *Landscape Research* 32 (2): 171–205. https://doi.org/10.1080/01426390701231564.

Zhao, Zoe. 2020. "Facing Down the Hong Kong Movement's Right-Wing Turn: A New Afterword." *Lausan* (blog). November 14, 2020. https://lausan.hk/2020/facing-down-the-hong-kong-movements-right-wing-turn-afterword/.

Zimmer, Alexis. 2020. "In a Fossil Fuel Economy, Air Can Become Unbreathable." In *Feral Atlas: The More-Than-Human Anthropocene*. Stanford, CA: Stanford University Press Digital. https://feralatlas.supdigital.org/poster/in-a-fossil-fuel-economy-air-can-become-unbreathable.

Zou, Hui. 2011. *A Jesuit Garden in Beijing and Early Modern Chinese Culture*. West Lafayette, IN.: Purdue University Press.

Zou, Xue-Yong, Zhou-Long Wang, Qing-Zhen Hao, Chun-Lai Zhang, Yu-Zhang Liu, and Guang-Rong Dong. 2001. "The Distribution of Velocity and Energy of Saltating Sand Grains in a Wind Tunnel." *Geomorphology* 36 (3–4): 155–65. https://doi.org/10.1016/S0169-555X(00)00038-6.

Index

Page numbers with *fig.* indicate figures.

Abe, Kōbō, 117

aeolian physics: desert research history, 250n52; dust events defined, 242n4; forestry as techno-ecological practice and, 56, 57; translation of, 251n1; *wind-sand* and, 10, 22, 243n12; wind tunnel construction, 43, 44, 45, 251n4. *See also* weather and weather systems

aerial seedings, 47, 51, 61, 63, 68, 70–71

aerosols: overview of, 8, 9, 10, 23; aerosol forms, 30, 32, 34, 250n49; COVID-19 and, 23, 262n2; environmental time and, 20, 245nn25,26; political formations and, 10, 20, 143, 146, 245nn25,26, 261n6; US and, 208–13, 265nn2,4

afforestation projects: overview of, 34–35; Convert Pastures to Forests mega-project, 78–79, 85, 255n3; downwind weather and, 87, 88, 91, 92, 93–94, 103; forestry bureau, 34–35; Great Green Wall, 34–35, 51–52; Korean Forestry Administration, 226; mega-projects, 34–35, 51–52, 78–79, 85, 255n3; Three-North Shelterbelt Project, 35–36, 51–52. *See also* forestry bureau; windbreak projects

agency: environmental governance, 101–2, 104, 105, 106; environmental subjectivity, 80, 90, 101–2, 104–7, 258n21; particulate matter in Beijing and, 37, 159–61, 165; representation in language, 3, 241n3

air, and gradients of dust: overview of, 37–38, 173, 179, 191–92, 250n51; bricks, 191, 196, 198–99, 198*fig.*, 264n10; capture and, 199–200; CCTV Tower, 189–91, 190*fig.*, 194, 195*fig.*, 196; data art project, 177; diamonds, 191, 195, 195*fig.*, 196, 198; intake devices, 179, 194, 198, 199–200; pollution-catching tower art project, 192–95, 193*fig.*, 195*fig.*, 196, 198, 199; vacuuming the sky art project, 191, 196, 197*fig.*, 198–99, 198*fig.*, 264n10

air conditioning, and airspacing techniques, 173, 178, 189

air design, and airspacing techniques, 180–81, 185, 187

air pollution, 8, 22, 23, 37, 57, 225, 247n32, 261n3. *See also* air, and gradients of dust; air conditioning, and airspacing techniques; air design, and airspacing techniques; air quality, and Beijing; air quality

human vacuum cleaner hashtags (*continued*)
    particulate matter; social media, and
    particulate matter
hwangsa (yellow dust), 10, 219, 222, 223*fig.*,
    225. *See also* downwind weather

infrastructures, and social and terrestrial
    engineering projects, 49, 55, 252n2,
    254n8
Ingold, Timothy, 247n33
Inner Mongolia: overview of, 50, 51, 57, 111;
    antidesertification programs, 34, 221,
    266n12; coal extraction, 57, 78; desertifi-
    cation, 33, 34, 87; dust events, 5, 214;
    medicinal plants and *rou congrong*, 82;
    Naiman County, 129, 130, 131, 133, 134,
    135; Qubqi Desert, 222, 223*fig.*, 227;
    Seoul-based tree-planting organizations,
    221, 266n12; Sino-Korean Friendship
    Forest, 219, 221-23, 223*fig.*, 225, 226-27,
    230. *See also* Alxa Plateau
inside/outside air, and airspacing techniques,
    180, 181-82
intake devices, and performance art, 179, 194,
    198, 199-200. *See also* air, and gradients
    of dust
international air pollution. *See* downwind
    weather; downwind weather, and United
    States
international relations: overview of, 211;
    ascent of China and, 8, 204, 206-7, 231-
    32, 238, 239; binational planting teams,
    227-28; dustflows and, 211, 214, 215, 230,
    232; friendship and, 226, 227, 228-29;
    futures and, 229-30; international cli-
    mate politics, 203-4, 207; Sino-Korean
    Friendship Forest, 216, 221-23, 223*fig.*,
    225, 226-27, 230; Sino-Korean relations,
    219-20, 221-22, 228; supply-chain capi-
    talism, 215, 266n9; *wind-sand* and, 220-
    21. *See also* friendship; geopolitics
International School of Beijing, 185, 264n8
interviews and fieldwork, 1-4, 33, 34
"Into the Air" (Calvillo), 177
iron, and oceanic iron cycles, 233-34, 266n2
isotopic analysis, 208, 211, 230-31, 232

Jameson, Fredric, 207, 248n36
Japan, 10, 214, 220, 224-25, 226, 265n8
Jianguo Xiongdi (Nut Brother), 196-99,
    197*fig.*, 198*fig.*
Johnson, Alix, 224

Johnson, Thomas R., 160, 261n3
Jullien, François, 21, 246n31

"kill goats, protect grass, and protect Beijing"
    (*sha yang, hu cao, bao Beijing*), 88, 93-94,
    256n11
Kirksey, Eben, 191, 192
Klein, Naomi, 257n14
Kohn, Eduardo, 259n8
Kohrman, Matthew, 186
Koolhaas, Rem, 189
Korean Forestry Administration, 221, 226,
    266n12
Korean Peninsula, and dust events, 10, 214.
    See also Seoul, South Korea; South Korea
Koselleck, Reinhart, 119
Kuriyama, Shigehisa, 187, 262n8
Kwon, Byung-Hyun, 219-22, 224-25, 226

lakes, and desertification, 111, 112, 119-20,
    258n2
land/territory: dust events and, 16, 244n18;
    historical ontology, 44-45, 251n5; political
    formations and, 16, 244nn18,19; tech-
    nopoetic process and, 52, 252n3; transfor-
    mations and, 9, 37, 45, 51, 124, 252n6. *See
    also* China
late socialism: overview of, 24-30, 247n35,
    248nn36,37,38,39,40,41, 249nn42,43;
    environmental governance, 26, 84,
    86-87, 249n43; open-endedness and, 13,
    17, 238; typologies of power, 15, 244n15;
    wind-sand and, 26, 27, 35, 249n43. *See
    also* environmental governance; phase
    shift; political formations; socialism
Lavauden, Louis, 245n22
lead isotopes, 211, 231, 232, 267n5
Lee, Bruce, 169
Lee, Li-Young, 41
Le Guin, Ursula K., 46, 252
Lévi-Strauss, Claude, 233, 251n4
Li, Yifei, 259n11
life expectancy-life/death, 144-46, 151, 163
Liu, Ken, 243n10
local specialty products (*tutechan*), 61, 100-
    101. *See also* sand commodities/products
Lock, Margaret, 161
long-haul truck driver jobs, 2, 65-66, 69
Lop Nur, Xinjiang region, 121, 121*fig.*, 123, 124
lungs: dust events and, 11; lung disease, 144,
    145-46; lung space, 187-88; particulate
    matter and, 142, 143, 144, 151, 154, 157, 173;

vacuuming the sky art project and, 199. *See also* breathing; breathing together

machine sky, 35, 50, 52, 53, 72, 250n51, 252n3. *See also* social and terrestrial engineering projects

Mao Zedong, and Maoism: adaptive governance, 25–26, 243n8; Foolish Old Man fable, 198–99, 264n10; futures and, 118; The People, 140, 141, 142, 143, 144, 164, 165; sheet of loose sand metaphor, 75; social-agricultural experiments, 51, 111; transformations and, 30, 37, 51, 252n6

Marx, Karl, and Marxism, 19, 237, 249n42, 259–60n11

Masco, Joseph, 124, 261n1

masks: airspacing techniques and, 188–89; COVID-19 in US, 261n4; types of, 180, 261n3; Xi as unmasked, 140–41, 140*fig.*, 142, 143, 144

materiality: desertification, 112–13, 118; ethnography(ies), 12, 244n14; land/territory and, 244n18; political formations and, 160, 237; *wind-sand* and, 10, 11

Mathews, Andrew S., 28, 171, 260n15

Mauss, Marcel, 160–61

medicinal plant market, and *rou congrong*: overview of, 100–101, 255n6; economic environment, 80–81, 84, 94; shrub growing and processing businesses, 77–78, 80–81; state buyers, 80, 82, 85–86, 99, 100–101, 102; state manipulation, 99–100, 101–2, 257nn17,18,19, 258n20; supply chain ecologies, 100, 102–4. See also *rou congrong*

mega-projects, and forestry bureau, 34–35, 51–52, 78–79, 85, 255n3

mercury fog, 234–235. *See also* weather and weather systems

metahistory, 115, 116, 135, 259n7

meteorology: overview of, 16–17, 22, 245n20, 246n28; Asia and, 222; dust events and, 7, 8, 220, 221; fixed forms versus phase shifts, 12, 244n13; friendship and, 226; geology and, 22, 71, 209, 232; international relations and, 229, 266n15; planetary atmosphere and, 13, 17, 22, 246n29; political formations, 13, 17, 216, 226, 245n21; stratigraphy of aerosols in US and, 208–13, 212*fig.* *See also* China's meteorological contemporary; weather and weather systems

Middleton, N. J., 242n4

Minqin County, Gansu Province: overview of, 111–12, 120*fig.*; antidesertification programs, 119, 120, 125; aversion of desertification model, 121, 121*fig.*, 122, 123–24; deserts as threat, 112, 113, 119, 120, 120*fig.*, 125, 260n13; futures, 113, 120, 125; Qingtu Lake, 111, 112, 119–20, 258n2; sandflow, 111–12, 113; vista/scene, 119, 122–23, 260n14; water reservoir, 119, 120*fig.*, 123

modern weather. *See* China's meteorological contemporary; weather and weather systems

modulation: economic environment, 87; engineering projects, 35, 49, 50, 54, 60; environmental governance, 100, 101, 105, 257n15

Mol, Annemarie, 119

monitors and monitoring conflicts, particulate, 9, 144, 146–47, 152–54, 155. *See also* air quality, and Beijing

monsters and weather-monsters, 234, 235, 267n4

mountain lions' health, 234, 235, 236, 267n5. *See also* downwind weather, and United States

Mount Tamalpais, San Francisco, California, 208, 210, 211, 265n2

Murphy, Michelle, 235, 244n14

Myers, Natasha, 54

Naiman County, Inner Mongolia, 129, 130, 131, 133, 134, 135

NASA: dust events, 6–7, 6*fig.*, 29; Pacific Dust Express, 31*map*, 38, 211, 216, 231, 265n7

necrosis, 143–46, 147, 165

neoliberalism, 66, 87, 256n9, 258n20, 264n1

Neyrat, Frédéric, 21, 249n43

Ng, Emily, 260n12

nightwind, 1–4, 2*fig*. *See also* wind; *wind-sand*

Ningxia Province, 77, 111

Nixon, Robert, 124

North America. *See* downwind weather, and United States

Nut Brother (Jianguo Xiongdi), 196–99, 197*fig.*, 198*fig.*

nylon sandbreaks, 125–26, 126*fig.*

oceans, 233–34, 236, 266n2. *See also* downwind weather

Olympic Summer Games in Beijing in 2008, 9, 150, 151, 152, 179, 205, 253n6

PM10, 151, 152, 153, 154–55, 156–57
poetics of particulate matter (poetics of dust
  events), 175–78
political collectivity: overview of, 144–45, 161,
  164–65; air and particulate matter, 143,
  145, 149, 160, 162–63; breathing together,
  141, 146, 147, 159–61, 162–63; necrosis
  and, 145, 147
political formations: aerosols and, 10, 20, 143,
  146, 245nn25,26, 261n6; becoming and,
  13–14, 16, 26, 237–38, 246n27; desertification and, 20, 124, 245n22; erosion/
  erosion-control measures, 61, 252n6,
  254n10; futures and, 118, 128–29, 135,
  259n10, 260nn12,17; groundwork and, 84;
  land/territory, 16, 244nn18,19; liberal civil
  society and, 146, 163–64; life and, 75,
  254n1; Marx and Marxism, 19, 237,
  249n42, 259–60n11; masks and, 141,
  261n4; materiality and, 160, 237; meteorology and, 13, 17, 216, 226, 245n21; open-
  endedness and, 12–13, 21; planet and,
  12–13, 14, 15, 20, 216; postsocialism, 24,
  258n20; scene of emergence, 19, 245n23;
  Sinocene, 207–8, 232, 238, 239; socialism,
  142–43; tear gas and microclimatic manipulation, 166–67, 168, 169–70, 171,
  262nn3,4; transformations and, 9, 20–21,
  37, 51, 246n27, 252n6; weather/weather
  systems, 12–13, 15–16, 20, 27, 245n24;
  wind-sand and, 15, 20–21, 245nn25,26. See
  also Chinese Communist Party; chronopolitics; environmental governance; geopolitics; governance of rural life; late
  socialism; Mao Zedong, and Maoism;
  political time; Reform and Opening; and
  specific countries, and leaders
political performance, and particulate matter:
  human vacuum cleaner hashtags, 161–63,
  164, 165; vacuuming the sky art project,
  191, 196, 197fig., 198–99, 198fig.; Xi, 140–
  41, 140fig., 142, 261n3
political time: futures and, 113–14, 124, 261n1;
  horizon of statist time and, 114–15, 116,
  117–18, 119, 121; sand and, 119–20; speed
  of, 118, 259n10; wind-sand and, 20,
  245nn25,26. See also political formations
pollution: overview of, 22, 23–24; air pollution, 8, 22, 23, 37, 57, 225, 247n32, 261n3;
  groundwater, 109, 110, 258n3. See also
  coal; fine particulate matter; particulate
  matter, and Beijing

postsocialism, 24, 258n20. See also late
  socialism
Povinelli, Elizabeth A., 133, 149
protests and protest movements: collectivity,
  170–71, 263n6; demonstrators becoming
  like water, 169–70; Hong Kong, 166–71,
  262n1, 263n5; Tiananmen Square, 169,
  196, 197fig., 262n2
purified air: air purifiers and air filters, 142,
  173–74, 186, 187, 188; canned air, 172–76,
  178, 188, 263nn1,3,4. See also airspacing
  techniques

Qian, Xuesen (Hsue-Shen Tsien), 254n12
Qicheng, Zhang, 178
Qingtu Lake, 111, 112, 119–20
Qubqi Desert, Inner Mongolia, China, 222,
  223fig., 227

Raffles, Hugh, 24, 242n6
RDIs (rotating DRUM impactors), 208, 209,
  210, 211, 212, 212fig., 213, 265nn2,3. See
  also downwind weather, and United
  States
Reform and Opening: overview of, 8, 206,
  246nn8,9; chronopolitics, 13, 14, 113, 118;
  deregulation, 89, 91; downwind weather,
  209; dust events, 8–9, 18; economic environment during, 90; Hong Kong, 167;
  lungs/lung disease, 144, 173; meteorology
  and, 17; modernity described, 19–20, 21;
  transformations and, 9, 51
resettlement programs for ex-herders: overview of, 1, 2fig., 18, 35, 36, 49, 51; aerial
  seedings, 51, 63, 68; antidesertification
  programs, 2, 3, 62–63, 241n2; economies,
  2, 64–65, 67–68, 69; governance of rural
  life and, 49, 51, 59, 63–64; grazing bans,
  53, 63, 64, 65, 68–69; home ownership
  and, 2fig., 3, 65, 66, 67, 69, 241n2; household registration on pasturelands and, 65,
  98, 254n13; open-endedness and, 66–67;
  precarity and, 64, 65–67, 69, 241n2; seed
  collection, 63, 68, 70, 71, 79; subsidies, 63,
  64; villages, 1, 2, 2fig., 51, 64, 65, 123;
  windbreak projects, 63, 64–65, 66, 67, 68.
  See also economies, and ex-herders; pastoral economy
retirement disbursements, 64–65, 90–91, 98.
  See also pastoral economy
Rheinberger, Hans-Jörg, 25
Rofel, Lisa, 13, 26, 118, 181, 248n41, 256n9

68, 69; environmental engineers, 50–51, 54–55, 253nn6,7; governance of rural life, 49, 50, 51; infrastructures and, 49, 55, 252n2, 254n8; machine sky and, 35, 50, 52, 53, 72, 250n51, 252n3; modulation, 35, 49, 50, 54, 60; open-endedness and, 66–67; sand commodities/products and, 61, 254n12; sand-control programs, 59–62; storm-sculpting, 56–59, 60–61, 62, 70; terraforming and, 53, 253nn4,5; textures of land and, 60, 62, 70, 239, 254n10; weather/weather systems, 47–48, 49–50; windbreak projects, 48*fig.*, 58, 60–61; *wind-sand*, 250n51. *See also* environmental governance; machine sky; resettlement programs for ex-herders
socialism: adaptive governance and, 25–26, 243n8; livable/nonlivable planet and, 205–6, 207, 264n1; particulate matter in Beijing and, 142–43, 145, 146; postsocialism, 24, 258n20. *See also* late socialism
social media, and Hong Kong protests, 170, 263n5
social media, and particulate matter: @BeijingAir Twitter feed, 150–51, 152, 153, 154; human vacuum cleaner hashtags, 161–63, 164, 165; tycoons and, 146–47, 155, 158; Weibo, 150–51, 153, 155–56
society/sociality: liberal civil society, 146, 163–64; particulate matter in Beijing and, 145, 147–49, 162–63, 261n7, 262n9; social stability, 63, 64, 66–67, 69, 92–93
Solanki, Aakash, 257n18
South Korea (Republic of Korea, ROK): Cold War era, 226; domestic air pollution, 225; factory pollution and, 221; Future Forest organization, 222–23, 223*fig.*, 225, 227; Korean Forestry Administration, 221, 226, 266n12; Sino-Korean Friendship Forest, 216, 221–23, 223*fig.*, 225, 226–27, 230; Sino-Korean relations, 219–20, 221–22, 228; tree planting projects, 227; yellow dust, 10, 219, 222, 223*fig.*, 225. *See also* Seoul, South Korea
Spirit of the Camel, 78, 81, 107, 255n2. *See also* camels
Spivak, Gayatri, 265n8
Stengers, Isabelle, 251n54
storm-sculpting, 56–59, 60–61, 62, 70. *See also* social and terrestrial engineering projects
Strathern, Marilyn, 250n50

straw: sandbreaks, 123, 125, 128, 134–35; windbreaks, 58, 59, 227–28
subsidies: forestry bureau, 92, 98, 254n13; grazing bans, 92, 98, 106, 254n13; resettlement programs for ex-herders, 63, 64; *suosuo* planting, 85, 98, 99, 100
Sun Yat-Sen, 73–76
suosuo (*saxoul*): overview of, 48*fig.*, 81, 82; camels, 77, 78, 107; desertification, 97; ex-herders as farmers, 81, 85, 95–98, 96*fig.*, 97*fig.*, 256n8; groundwork, 83–84, 99; holoparasitism and, 77–78, 82, 94; sand commodities/products and, 81, 85, 94; subsidies for planting, 85, 98, 99, 100; *suosuo, rou congrong*, and ex-herders triangle, 91–95; windbreak projects, 79, 81, 82, 83, 85, 255n4. *See also rou congrong*
supply chain issues, 100, 102–4, 215, 266n9
swallowing (*tunshi*), 120, 260n13

tax breaks, and environmental governance, 79–80, 100
tear gas, and microclimates, 166–67, 168, 169–70, 171, 262nn3,4
techno-ecological practices, 56, 57, 250n52. *See also* forestry bureau
technopoetic process, 52, 252n3
Tengger Desert, 2, 64, 260n13
textures of land, and engineering projects, 60, 62, 70, 239, 254n10
time, 20, 245nn25,26. *See also* environmental time, and deserts; environmental time, and futures; political time
traditional Chinese medicine, 77, 79, 82, 87, 187
trees: forestry bureau windbreak projects, 1, 85, 255n7; Future Forest organization, 222–23, 223*fig.*, 225, 227; Korean Forestry Administration, 221, 226, 266n12; Sino-Korean Friendship Forest, 216, 221–23, 223*fig.*, 225, 226–27, 230; tree-planting organizations in Seoul, 221, 266n12. *See also* forestry bureau
Tsing, Anna Lowenhaupt, 245n20, 260n15
Tuan, Yi-Fu, 259n9
tycoons, and social media, 146–47, 155, 158

United Nations Conference to Combat Desertification, 226
United States (US): Asian relations and, 265n8; Clean Air Act, 210, 231; Cold War era, 124, 265–66n8; color-coded threat

green initiatives, 249n42; international climate politics, 203, 204, 207; masks and, 140–41, 140*fig.*, 142, 143, 144; outdoor visit during air quality alert, 139–41, 140*fig.*, 261nn1,3,4
Xinjiang region, and Lop Nur, 121, 121*fig.*, 123, 124

yellow dust (*hwangsa*), 10, 219, 222, 223*fig.*, 225. *See also* downwind weather; dustflows

Zapruder, Matthew, 225
Zhan, Mei, 13
Zhu Rongji, 15–17, 87, 88

Founded in 1893,
UNIVERSITY OF CALIFORNIA PRESS
publishes bold, progressive books and journals
on topics in the arts, humanities, social sciences,
and natural sciences—with a focus on social
justice issues—that inspire thought and action
among readers worldwide.

The UC PRESS FOUNDATION
raises funds to uphold the press's vital role
as an independent, nonprofit publisher, and
receives philanthropic support from a wide
range of individuals and institutions—and from
committed readers like you. To learn more, visit
ucpress.edu/supportus.

www.ingramcontent.com/pod-product-compliance
Lightning Source LLC
Chambersburg PA
CBHW020823270326
41928CB00006B/421